Praise for Richard Manning's

"Readers of [Manning's] earlier books—~ ~
Round River, and *Grassland*—will find h ~
background information, the same passio~~~~ ~~~~~~~ ~~~ ~~~~~~-
ical worldview, the same deft weaving of personal experience into a
larger story . . . The ultimate goal of all these efforts is to create a permaculture—a way, not merely of feeding ourselves, but an entire way
of living that might be sustained, with joy and dignity, as far as we can
see into the future. *Against the Grain* shows us how difficult that task
will be, in light of our agricultural history, and how vital."
—Scott Russell Sanders, *Orion*

"A provocative and engaging read."
—Lynda V. Mapes, *The Seattle Times*

"[A] wide-ranging critique of the role of agriculture in human society
. . . [*Against the Grain*] is aimed at provoking . . . and in this Manning
succeeds tremendously, expanding our consciousness of where we are
and how we got there, and opening up a debate for years to come."
—David Barker, *The Santa Fe New Mexican*

"An exhilarating and provocative questioning of our most ingrained be-
liefs about how we get our food and why. A must-read for anyone concerned about the intimate couplings of man, plant, and beast."
—Betty Fussell, author of *The Story of Corn*

"*Against the Grain* is a brilliant, provocative book. Where environmental journalism is concerned, Richard Manning is at the head of the
class."
—Larry McMurtry

"Controversial and prodigiously researched . . . Manning skillfully details the historical spread of agriculture."
—*Publishers Weekly*

"Manning brings theory to life with well-crafted essays that cover such
diverse subjects as the Irish potato famine and the controversy over
bioengineered plants. Readable and well-researched, this book unset-
tles as it informs."
—Patricia Monaghan, *Booklist*

© Tracy Stone-Manning

RICHARD MANNING

AGAINST THE GRAIN

Richard Manning writes regularly about the social, political, and ecological threats to America's West and has been called "a gifted environmentalist with a refreshing sense of humor" by *The New York Times*. He's been awarded numerous honors, including the Richard J. Margolis Award for promising new journalists reporting on social justice, the Montana Audubon Society Award for environmental reporting, and the C. B. Blethen Memorial Award for investigative journalism. His other books include *Last Stand*, an examination of the logging industry; *A Good House*, a lyrical account of building a log cabin in the Montana wilderness; *Grassland*, an exploration of the destruction and recovery of the prairie ecosystem; and *Food's Frontier* (North Point Press, 2000), an eye-opening investigation into the future of the world's food supply. His book *One Round River*, a moving depiction of the slow but inexorable degradation of the Blackfoot River in Montana, was awarded a Pacific Northwest Booksellers Association Book Award in 1999. He lives in Montana.

ALSO BY RICHARD MANNING

Last Stand

A Good House

Grassland

One Round River

Food's Frontier

Inside Passage

AGAINST THE GRAIN

HOW AGRICULTURE HAS HIJACKED CIVILIZATION

RICHARD MANNING

NORTH POINT PRESS

A DIVISION OF FARRAR, STRAUS AND GIROUX

NEW YORK

North Point Press
A division of Farrar, Straus and Giroux
19 Union Square West, New York 10003

Copyright © 2004 by Richard Manning
All rights reserved
Distributed in Canada by Douglas & McIntyre Ltd.
Printed in the United States of America
Published in 2004 by North Point Press
First paperback edition, 2005

The Library of Congress has cataloged the hardcover edition as follows:
Manning, Richard, 1951–
    Against the grain : how agriculture has hijacked civilization / Richard
Manning.— 1st ed.
        p.   cm.
    Includes bibliographical references and index.
    ISBN 0-86547-622-5
    1. Agriculture—History.   2. Agricultural systems—History.
3. Agriculture—Social aspects—History.   I. Title.

S419.M26 2003
630'.9—dc21

                                                              2003013718

            Paperback ISBN-13: 978-0-86547-713-1
            Paperback ISBN-10: 0-86547-713-2

            Designed by Jonathan D. Lippincott

                    www.fsgbooks.com

            1   3   5   7   9   10   8   6   4   2

CONTENTS

AGAINST THE GRAIN

AROUSAL

It is high summer at my mountainside home in Montana, when days are long at this latitude. The season is, if not yet desperate, at least frenzied, because we who live in the northern Rockies know that the climate makes life difficult when winter comes. The cold comes fast, so flora and fauna grab for photosynthesis or its various derivatives while they can, in a constant buzz and bustle. The observer of all this sees a race for a more prominent seat in the sun: set seed sooner, get taller faster to shade a rival, learn to grow in a bare spot no other can tolerate, secrete poisons to kill the competition, send out roots to steal water that belongs to others. The color that traces this race is green; primary producers announce their success in sucking up sun by making a display of it. The various subplots thickening among the wider cast of characters spin off the main plot played out among plants. Everything depends on the plants' success, on green. All the rest is secondary.

Mimicry, attraction, repulsion, information. These secondary dramas are a primal struggle staged in a blinding array—the red of devil's paintbrush, the yellow of arrowleaf balsamroot, the blue of pentstemon, the gaudy flash of butterflies drawn, like me, to these colors. Too often they draw me away from the keyboard, into the hills, into the vital stories of others. Every shade signals something about food and sex, the links in every being's path to the future. Unlike me, though, the butterfly is a player on this stage. To my eyes, this is simply a pleasant display. My pleasure is vestigial, relict, when compared to the intensity with which it is examined by all those other eyes around me. For them, reading the code of color is survival.

Color reveals a layered natural text, most of it straightforward and necessary information. Some of it is the ink on a contract of partnership: the black sheen of chokecherry or currant signals to bears and birds that the fruit is ripe, and then chokecherry sprouts from bear shit and bird shit miles away. This is how chokecherry uses color to recruit bears and birds to spread its genes. Sometimes, there is deceit, as when a perfectly edible butterfly evolves into a dead ringer for a foul-tasting species. There is no premium on honesty or fair play where survival is concerned.

Every so often I test this observation with a cruel trick, leaving the bright-red plastic jug that holds gas for my chainsaw out on my front porch. On the right sort of day, calliope hummingbirds arrive within minutes to buzz and hover an inch or so away, believing they have encountered the world's biggest flower. I don't mean to lie to them. More, I am a child trying to form the first few letters of a fully formed and unimaginably complex alphabet of hummingbirds. Color makes me want to be able to read what is being said all around me. It is not a trivial desire, more an ache, a longing that grows out of what I suspect is a profound loss—even more lost to many of my fellow humans who don't live as I do, with long days to think and wander a relatively undisturbed mountainside forest.

Every fall I try to plumb some of the depths of this longing with a sort of game some humans play. I hunt meat, an experience that is a pathetic substitute for what the hummingbird knows and sees every day, but one of the few ways I have of sustaining a life that is visibly coupled to the forces that created it. I don't kill for sport but for meat. Venison and elk provide most of the protein at my wife's and my table. Through the years, I have come to look forward to these fall weeks of hunting, less for the killing than for the seeing. Then I am able to perceive more detail—literally. Hidden layers become visible. Every subtle shift of shape or sound in the forest may signal the advent of an important moment. Hunting enlivens the senses like no other experience, giving me a taste of what it must be like to truly see and hear.

Sometimes I hunt birds with dogs, and, again, the pleasure is not so much in killing the birds as in observing the dogs, equipped as they are with noses many times more powerful than mine. They raise their heads and face the wind, and their nostrils pulse and flare as if they are pumping in every bit of air to sample the surroundings. They smell the world in the same way that I see it.

I came to believe it was possible for me to know more of the world this way. One year I was hunting elk, and before seeing or hearing the beasts, I smelled them. I could smell the coming kill, and I was right. We know more than we think we do. Experiments have shown that normal people given the whiff of a T-shirt can tell whether it was worn by a man or woman, even have a sense of whether the wearer was attractive. How much of this have we sublimated? What would it be like to meet other humans and smell, say, anger, or arousal, as I am sure my dogs do?

Sensual experiences of this sort have spawned a certain curiosity in a few people, have sent perfectly civilized and privileged modern individuals to live among primitives in an attempt to comprehend in some blundering, bumbling manner just where the doors of our perception open. Richard Nelson, an anthropologist and a friend, made such a pilgrimage, spending his early adult years living among the Koyukon people of the Yukon River basin, an experience he recounts in fine books like *Make Prayers to the Raven*.

Nelson once told me a story. He had left the Yukon and settled on an island in the far different maritime environment of southeast Alaska, where he spent years hunting deer and closely observing the place, about which he eventually wrote *The Island Within*. He still maintained contact with his Koyukon friends, and after several years he invited a couple of them to come visit him—a big deal for them, in that they had never been outside their familiar surroundings. Geographically unmoored as we postmoderns are, it is difficult for us to imagine what this means, but try to understand a human life in which all referents, all mental anchors and guides, are set in the natural features of a particular place.

Nelson, of course, was curious as to how they would respond, but not quite prepared for the way this reunion with old friends worked out. It was silent. As with all reunions among friends, there was catching up to do, greetings to be made, stories to tell—but none of this happened, at least not right away. His friends simply lapsed into silence. They were so wrapped up in observing everything around them that there was no part of their brains left for speech. This went on for days. When finally they did speak, they revealed to Nelson all sorts of details about his closely observed island that he himself had never noticed.

There are beings, many of them human beings, that see, smell, hear, remember, sense more than we do. This is not a genetic accident, like being taller than six-foot-five or having an IQ of 150 or high cheekbones. This is a matter of culture. The human beings who maintain these hyper-refined senses are hunter-gatherers. Their impressive powers of perception have been noted and detailed by just about every student of hunter-gatherer groups. It is not only that they sense more than the rest of us do, but that they do so in a qualitatively different fashion. In his book *The Spell of the Sensuous*, David Abram leans on philosopher Maurice Merleau-Ponty's concept of synaesthesia to explain Abram's own experience with hunter-gatherer perceptions. The term "synaesthesia" describes something every child knows. In fact, Merleau-Ponty believes that we have "unlearned how to see, hear, and, generally speaking, to feel." Synaesthesia is the mental function (or suite of functions) in which the senses run together, in which colors have a feel to them and tastes have a color. We speak of a loud shirt, of bright music, yet how often do we sense reality this way? For Abram and other observers, the phenomenon marks a total immersion in sense, when the observer is no longer in control, no longer separating and analyzing sight, sound, and texture, and becomes a part of his sensual surroundings. That is, the observer calls forth the world.

"As soon as I attempt to distinguish the share of any one sense from that of the others, I inevitably sever the full participation of my sensing body with the sensuous terrain," writes Abram. "Many indigenous peoples construe awareness, or 'mind,' not as a power that resides inside their heads, but rather as a quality that they themselves *are inside of* along with the other animals and plants, the mountains and the clouds."

I mention all of this because my own hunting tells me it is true, and that is part of why the experience enlivens me. Yet I know it better still from the more common experience of making music. I became a musician as an adult. The advantage of that approach is that one has an adult's drive and discipline to pound away at practicing the necessary technical skills day after day. The disadvantage is that one needs the drive and discipline. Merleau-Ponty is right about "unlearning." As adults, we have unlearned how to hear. As a literate society, we trust notes on a printed page—indeed, authorities of all sorts—far more than we trust our own ears, such that a large part of my struggle to learn music has been to teach myself to hear. Immerse a child in music before this unlearning has occurred and real music, not just notes, flies from her fingers like the sung notes of a bird.

Yet lately, after years of learning the fundamentals and years of learning that the biggest barriers I face are those I unconsciously erect for myself, I have begun to believe what a music teacher once told me: "My main job is to teach you that you know a lot more than you think you know." The physical marker of this dawning awareness has been for me something very much like what Abram describes. I finally began hearing chord changes, and hearing a C as a C, only when the senses blurred, when individual notes took on color and texture, not just sound. There came a day when I realized that some notes had edges, had thickness, and I found myself shifting position to try to see the backside of a note, so real was the sense of depth. At the same time, I found that more and more it was not my fingers and mind playing the music but my whole body. For me, this is not about making music; it is about exploring the horizons of what it means—or at least meant—to be human.

There was a time in my life when I enjoyed fly-fishing, before Montana's rivers became a floating fashion show for the catalogue-fresh Orvis gear of dot-com millionaires. I can understand, though, why those with means gravitate toward this pursuit; it's not just about catching fish. No other form of hunting I know requires such a thorough knowledge of the prey's habits and surroundings. Fly-fishing is simply a means for contemplating rivers.

Once I was fishing a favorite stretch of a favorite creek at the finest hour of the day. Dusk was thickening to dark on a late summer evening and all the right bugs were on the water. I was taking fish and really didn't want to quit, but the fact was, I could no longer see. One can still fish under such conditions if the mind can correctly fill in the blanks where vision fails, but mostly the mind guesses and misses. Nonetheless, I played along. All of a sudden, my mind's eye leapt from its usual perspective to zoom in on a scene across the creek where I imagined my fly to be drifting toward a productive hole. I saw that particular spot as if in close-up, but also fully illuminated, bright as day. Then, just as my imagination had provided daylight, it delivered an enormous rainbow trout, arching out of the stream to take my fly. I felt a little stupid when this hallucination seduced me into tensing my rod to set my hook in the phantom fish. Less stupid when the hook set solidly, and I fought a nice fish for a while, finally landing it. It was the fish I had imagined, no mistaking it.

This is a book not just about agriculture but about the fundamental dehumanization that occurred with agriculture. It will argue that most of humanity struck a bitter bargain over the past ten thousand years, trading in a large measure of our sensual lives for the bit of security that comes with agriculture.

We can't really conceive of what we humans lost with the process of civilization—with agriculture—until we ask what human nature is. What makes us human? Our intelligence? Our use of tools? That we are oversexed? Social? Tinkering? Bellicose? Self-aware? Each of these traits has been put forward at some point to distinguish us from the rest of the animal world, and, to a degree, that discrimination is the point of the exercise. We manufacture dichotomous answers to the question, us versus them, because, at least unconsciously, all too often consciously, the answer justifies our dominion, our shabby treatment of the rest of life on the planet. When we examine these attributes, however, we find little reason for dichotomy. Crows use tools, as do chimps. Gorillas are self-aware. They pass the mirror test—that is, they know the image in the mirror is a reflection, not another ape. So do chimps and dolphins. Porcupines are oversexed. (Yes, I know: *very carefully*.) So what makes us unique?

The search for an answer has sent a lot of thinkers to places like Lascaux, in the south of France, where in the caves, as elsewhere in Europe, are accomplished paintings of animals, the result of fine observation. They flow from a sense of wonder, and a need to preserve that aesthetic experience in art, an impulse felt as long as forty thousand years ago. Anthropologist Ian Tattersall invokes the impression they give at the beginning of his book *Becoming Human*, where he describes the thirteen-thousand-year-old drawings at the Combarelles caves in France:

> You are mesmerized, not simply by the subtlety of these marvelous engravings—done at the time when a landscape around Combarelles, now oak forest, was open steppe roamed by mammoths, wooly rhinoceroses and cave lions—but by their sheer ancientness. For this is not in any sense crude art; it is art as refined in its own way—and certainly as powerful as anything achieved since.

In other words, these people were like us, and therefore worthy of our contemplation. Art is evidence of their humanity. For some, the European caves make a particularly comfortable beginning point for this rumination, because their inhabitants were "European," and thus presumably in our lineage—"our" meaning the branch of Caucasians that eventually colonized the world and became the dominant culture. Leave aside for the moment that this short-changes the other geographical branches of the human endeavor. If there is such a thing as a human nature, it doesn't really matter much which set of humans we examine first. Europeans will serve as well as any others.

The painters of Lascaux, however, were, in a very real sense, not "us." They were Cro-Magnon people, and while it is generally held that the European lineage descended from them, I'll argue later that this is simply not true, neither culturally nor genetically. The ancestors of the Caucasian line arose elsewhere. The Cro-Magnons were hunter-gatherers, of course. In their time, there were no agricultural people. Yet does their art make them like us? Certainly it does, and certainly it is a marvel. Does it define the beginning of a human nature? Not really, in that I think it is derivative, as are those previously cited defining traits like sexiness, tool use, and self-awareness. I am with Tattersall in my wonder at what appears on these walls, but I think the art is the result of a far more dramatic and fateful turn of events, an accident of evolution, hundreds of thousands of years before. It was that accident that made us human.

The highly touted human brain is but half of the chicken-or-egg story of what makes us who we are. The other half is suggested by the old adage "You are what you eat" (a pun in its original German form: "Mann ist was mann isst"). As is the case with chickens and eggs, it is pointless to ask which came first, our appetite or our brain, in that each depends intimately on the other.

The clear physical evidence of this quantum leap in hominid evolution is the enlarged cranial cavity that showed up suddenly in our lineage's fossil record about 2.5 million years ago. That is, in an evolutionary instant, a strain of upright apes became equipped with a larger brain, about half again as large, proportionate to body weight, as the brain of other primates. Besides its sheer size, there is evidence that the brain's configuration—its circuitry, if you will—became more sophisticated in the bargain. The planet has not been the same since.

I have argued above that all of those attributes that we believe set humans apart from the rest of creation are derivative, but this development of the brain is primal, axiomatic. That is, a quick change in brain size is a genetic quirk, an accident, a mutation, and that accident might have occurred in most any species with a brain. That's not the interesting question. More interesting is: Would it be a good thing? In evolutionary terms, would it have conferred fitness if it had occurred in some other species? So highly do we regard our big brain that the question seems odd, but imagine for a moment a hyper-charged brain not imbedded in a body as mobile as ours, or, more important, in one suited, with its unique hand, to make wide use of tools. Super-intelligence still might be an advantage to such a being, but would it be enough of an advantage to overcome its cost?

A brain, energy-wise, is an enormously costly organ. That is, pound for pound, it takes far more energy to keep it running than any other organ. It accounts for about 3 percent of our body's mass, yet close to 20 percent of our energy needs. The quantum leap in brain size brought with it a stiff increase in the individual's need for calories. Thus, if that bigger brain did not bring with it at least a pro-portional leap in our ability to feed it, there would be no net gain, and it would not confer fitness, would not survive evolution's merci-less test.

It was in this sense that the development of a big brain was in fact derivative, in that our line of apes already had some assets at work. The sudden leap in brain technology made a bigger, flashier engine in a chassis that was already engineered to handle it. We know this because our closest relatives that didn't get this big brain are still fairly impressive animals, relying on a set of skills that would dovetail nicely with souped-up intelligence. Our nearest relative is the chimpanzee, of which there are two species. In fact, an honest taxonomy—if such a thing is possible, with our anthropocentric bias—would argue, as some scientists have, that chimps comprise three species, and we are the third.

We can safely assume, however, that the fundamentals of chimp existence common to human existence were present in a com-mon ancestor. Shared behavior argues for this. Chimps are, for in-stance, highly social. They are unrelentingly libidinous, especially the smaller of the two species, the bonobo (or pygmy chimp). They have a division of labor. Males hunt meat in groups, with organized and complex strategies. Their primary prey are colobus monkeys and young antelope, but they hunt as many as twenty mammal species, not to mention some birds.

More important, though, their hunting figures into a broadly omnivorous lifestyle. They are generalists, relying on many different plants and animals to support themselves. I once got a sense for how far this goes when walking in a stream-slit canyon in western Uganda with a caretaker of the protected habitat for a band of chimps. The guide pointed out one particular species of tree and told me that whenever the chimps get diarrhea, they visit that tree and eat its leaves. Analysis of those leaves showed they contain a substance very effective in suppressing diarrhea.

It is this already developed generalist's strategy that allowed our big brains to be a gain in terms of fitness. We could use them to store and sort our expanded catalogue of food.

The argument that these are interdependent developments is supported by a haunting example cited by Susan Allport in her book *The Primal Feast*. The koala of Australia, cuddly as it appears, is a notoriously dimwitted beast. Further, it is almost wholly dependent for its food on a single species, the eucalyptus tree. Not so long ago, in evolutionary terms, however, the koala ate a large array of plants. Dissecting a modern koala reveals a tiny brain rattling around in a large brainpan. Evolution has wiped out the need for a large brain, so presumably downsized it to avoid the energy costs, but evolution has not yet had time to downsize the cranial cavity.

There are two broad lines of nutritional strategies among primates. The slothlike, small-brained sedentary line eats mostly leaves, a low-quality, high-volume food patiently digested in a large (and also energetically costly) digestive tract. The larger brain dictates a wider, more energy-rich diet, which in turn dictates motion. It also comes with a shorter digestive tract, which, in part, compensates for the energy costs of the brain but also requires high-quality food. Anthropologists speculate that humans came into this set of equipment just as the conditions of life in Africa were changing, such that this new strategy would confer fitness.

The environment was becoming drier; forests, those great leafy masses of fodder for the sedentary, were disappearing, being replaced by grasslands and large, grazing herbivores. Just as we developed the need to hunt more and the brain to allow us to do so, there appeared a large supply of animals worth hunting—big, rapidly reproducing piles of meat. Predecessor primates likely had eaten meat, but now conditions aligned for them to depend more heavily on it.

Meat protein, however, would not do the job alone, because meat lacks some necessary nutrients and is an inefficient way to satisfy all of a body's carbohydrate needs. So we also began using our big brains to identify and locate plants that stored carbohydrates in starchy taproots. These supplemented the range of fruits primates had long consumed. We can imagine the shape of these changes on a variety of levels, chief among them the newfound priority on information. This was the beginning of the information age. Humans began satisfying their increased appetites by consuming an increased number of species, both plant and animal. This required more than the knowledge of antelope and colobus monkeys that had sustained primates to this point. New beasts with new habits became prey. Human hunters were forced into competition with bigger, more dangerous predators, with the result that humans themselves became targets, requiring new strategies for defense.

The new demands on the plant side were even more daunting, and likely placed the extra burden on women. There are a range of arguments as to why a division of labor makes sense; most are linked to the fact that plants are stationary and so dovetail with the duties of child rearing. Whatever the reason—those offered are speculation—we do know from the examples of surviving hunter-gatherer societies that a detailed knowledge of plants is usually a specialty of women. What fruit is ripe when and where? How best to collect and store it? What plants satisfy which needs in what seasons? Buried in these questions is what the psychologist Paul Rozin of the University of Pennsylvania labeled the "omnivore's dilemma." That is, the increased appetites of the omnivore lead to using an ever broader array of species, which leads to experimenting. The real gain from that novelty is a broader base of support for the species. When eucalyptus trees all die in a given place, so do all the koalas, but omnivores have options. The cost of that experimentation, however, is learning the hard way which plants are poisonous, and which parts of plants are poisonous, or poisonous in certain seasons. The legacy of exploration remains in humans' routine consumption of toxic plants. The rest of the animal world can't tolerate the bitter tannins that we relish in our teas. Potatoes are full of nasty alkaloids that we render harmless by cooking. The knowledge must have come at a terrible cost. Some mistakes were fatal.

All the more reason to grant importance to information, to cataloguing and cross-referencing it in that extra space in our cranial cavity, but above all, to gathering it. Our link to this information was and is our senses. Our survival as a species depended on a keen sense of sight, especially in using color cues to draw us to fruit, to discern subtle variations that distinguished an edible fruit from a poisonous look-alike, to see and smell which leaf above ground signaled the presence of a ripe taproot below. Hunters must have sat for hours analyzing every muscle twitch of prey. Every snap and crack in the forest carried a message, every shift in the wind a shift in fortunes.

There must have been a huge role for synaesthesia in all of this. We could go further still and suggest, as Abram did, that the hypersenses of humanity create our world—that we see, touch, and smell it into existence. This is the notion that may help explain to a rationalist's understanding the Australian aborigines' concept of Dreamtime. There is a parallel universe, which is Dreamtime; our senses call it into existence and make it into the real world in which we live. But we don't need to go so far down this epistemological path to understand that this fundamental change in brain and appetite sponsored the changes that made us human.

Part of what distinguishes humans from all other species is their ubiquity. Before the quantum leap in brain size 2.5 million years ago, there were a number of upright primates, using tools, socially organized, coexisting in Africa. With their merely adequate brains, all of these primates had a limited range; with bigger brains, that limit lifted and the great diaspora began. Yet we have no reason to believe that a population explosion drove our species out of Africa. Instead, our omnivorous appetites gave us command over an ever-wider diet, which in itself granted us the ability to range over the planet, and to weather periods of upheaval like the ice ages. As of 300,000 years ago, the human brain had evolved to roughly its modern size. As of 100,000 years ago, humans had inhabited a part of the Middle East and Asia. We were in Europe 40,000 years ago. We may have been in Australia as far back as 60,000 years ago, but certainly by 45,000 years ago; Siberia shortly thereafter; North and South America no less than 15,000 years ago. No other species is so widespread. We achieved this range without agriculture, with nothing more than stone spears and all the information we could hold. Information was gathered by, even spread by, another tool of ours: social cohesiveness. Food bound us together.

During the roughly 290,000 years from our beginnings as humans to the beginnings of agriculture, there was one long, slow trend in our existence. Early humans consumed a few plants and a few very large prey animals, then gradually shifted to smaller animals and a wider variety of both plants and animals. This trend stands out in the archaeological record, but we can appreciate the logic of it by examining the Plains Indians of North America and their hunting practices before Europeans came. The widely used technique then—and the evidence says this technique stretches back to the beginnings of humans in North America—was the "buffalo jump." A group of people, men and women alike, used their knowledge of buffalo movements to haze and herd the animals toward a cliff, then stampeded them over the edge. The payoff in calories for this minimal effort was enormous, but it required social organization and sharing, both of the labor involved and of the proceeds.

No single human, or even a family of humans, can eat a buffalo or a mammoth in one sitting. There was no way to store the meat, no way for an individual to accumulate wealth, so communal feasting became the payoff for social organization. The primate line has long held with this practice. Male chimps gather for ritual feasting. Dominant males share choice bits of food with subordinates to reinforce the social structure, the proto–business lunch. Chimps are even more adept than we are at using food to forge sexual bonds. When a group of bonobos finds a large tree full of ripe fruit, they don't immediately attack it, but instead celebrate by everyone copulating with everyone else. Male bonobos routinely trade meat for sexual favors after a successful hunt, the proto–dinner date.

The difference between chimps and humans in this regard is promiscuity. The norm in chimp societies is to stage riotous orgies of public copulation. The norm in human societies is private copulation and lasting (or serial) bonds, either monogamous or polygamous. Of course there are examples of human societies more promiscuous than others, but, interestingly, anthropologists have identified a strong positive correlation between promiscuity among humans and males' time spent on, and attention to, hunting. There is a long list of genetically encoded reasons as to why this strategy makes sense. For instance, a few males, with luck, kill a large amount of meat, enough to go beyond their immediate families' needs. Trading it for sex is a way of extending the "family" to better allocate the meat. At bottom, though, is the fact that our big cranial cavities require young to be born in a relatively primitive and helpless state. All of this implies a division of labor, a marital contract written in terms of food.

The power of food to bond humans together can also be seen in a counterexample, reported by Colin Turnbull in the late 1960s, when he was curator of African ethnology at the American Museum of Natural History in New York. Turnbull spent the years 1964–67 living among the Ik people in Uganda, who, because of a political upheaval, were forced from their native hunting lands to a barren mountainside. Simply put, they had no food, so of course there was no food sharing; but all of what we moderns and even our hunter-gatherer ancestors regard as the fundamental decencies of the human condition fell apart: children stole food from old people; one mother rejoiced when her child was snatched and eaten by a leopard; children were abandoned. All family structure broke apart. Allport comments on Turnbull's report: "The Ik had no place else to go, and they abandoned instead many of the things we take for granted about being human, *many of the things we never thought could be abandoned*: love, cooperation, empathy, the sharing of food." (Emphasis in original.)

Food makes human bonds possible. The anthropologist Lorna

Marshall reported in detail how this binding works among Africa's !Kung bushmen with the sharing of meat immediately after a kill:

> The fear of hunger is mitigated; the person one shares with will share in turn when he gets meat and people are sustained by a web of mutual obligation. If there is hunger, it is commonly shared. There are no distinct haves and have-nots. One is not alone. . . . The idea of eating alone and not sharing is shocking to the !Kung.

Big brain aside, we still are animals bound by the primary law of the animal world—namely, that our attentiveness first and foremost is focused on food and sex, because these are the prime guarantors of the survival of our genes. Our larger brain does not exempt us from this requirement; it simply shapes the ways in which we satisfy it. It creates in us an attentiveness to the conditions of life so intense it borders on love. Some have called it that. The biologist E. O. Wilson raises the notion of biophilia (in his book by the same name), literally, a love of life. What he means by this is an attentiveness through our senses of the conditions of life. He argues that such an intense devotion to our surroundings would indeed confer fitness. That is, it would provoke an obsessive focus of one's senses on gathering the information necessary to ensure our survival and the survival of our genes.

Think now of a red, ripe plum. I react first to its deep color, a wine red that can grab the eye from across the room, shining through a maze of all other colors. How long has this very shade of red been drawing human eyes? Think next of its shape, the round, full shape that provokes synaesthesia, a blur now to the (in my male mind) feminine butt. How quickly this red and this curve can cross from senses to sensuality. How often do these colors and these lines, from Lascaux and Combarelles forward, re-create themselves in our art, in the expression of our sense of the aesthetic that gave rise to painting, dance, and music, even at the very beginning? We can ask the same question about this art as we can about the human brain: Why is there art? What fitness does it confer? Why do we find it so deeply satisfying? Does this aesthetic, wrapped as it is in food and sex, perform something of the same function that food does, the function so glaringly absent among the displaced Ik?

People who study and live among the vestigial hunter-gatherers of the world usually develop affection and respect for them, set in a wistfulness for what has been lost. I have dwelt much on the nature of their economy here, but those who have spent time with them more often mention their distinctive character traits such as honesty, affection, and humility, or social traits like egalitarianism. Such societies are generally without leaders or hierarchy, without rich and poor. Either all enjoyed abundance or all suffered want. These attributes are no accident, as comparisons with early agricultural societies will show us later, but this wistfulness for the hunter-gatherer life needs some tempering, lest we fall prey to stereotypes of the "noble savage" sort. Hunter-gatherer life could be as violent as our own when tribe encountered tribe. Indeed, even chimpanzee males from a given group form gangs to attack and kill the males in a neighboring group. Infanticide and cannibalism are both common among hunter-gatherers (as they also have been among "civilized" agricultural people).

No matter what the hunter-gatherer character, however, it is important that we try to understand it, because in doing so we begin to define human nature. If there is an inherent nature in humans, it is a product of evolution. Evolutionary pressures take a long time to enter and to leave our genome. Our kind has spent at least 290,000 years as hunter-gatherers, only 10,000 as agricultural people, making the latter way of living a relatively brief and novel experiment. Only small traces of agricultural life can be read in our genes. We still run on hunter-gatherer software.

Our species spent its formative years in Africa, largely, we imagine, because there was no real need to move on; nor had we yet evolved our bag of tricks sufficiently to give us the generalist's ability to do so. There was no need because hunter-gatherer populations are generally stable. Being in constant motion, they have no soft foods to replace mother's milk, so children are not weaned until they are about four years old. The resulting spacing of births, about four years apart, along with a measure of infant mortality and infanticide, keeps populations static.

Just as likely, we had some enemies back then, large predators that kept us in check, just as tigers and mountain lions prey on humans today. Much is made of the deep-seated preference of humans for a room with a view, a commanding vantage that allows us to survey a large area, the reason why trophy homes are built on hillsides and apartments with a view of the park demand a premium price today. It's often suggested that we get this preference from our grassland days: we needed a commanding view to avoid being eaten.

Thus our apprenticeship taught us not only to use tools, to eat a wide variety of foods, and to hunt, but also to avoid predators. We left Africa, spread to the Middle East, Europe, and Asia, and then took these tools to Australia and finally across Beringia (the former land bridge, now the Bering Strait) to the New World. On the two most recent stages of that trip, where the evidence of our coming is clearer, the archaeological record shows a massive wave of extinctions. In North America, for instance, all of the large mammals indigenous to the continent became extinct, including wooly mammoths, giant bears, tigers, sloths, camels, and horses. The big mammals we have left, such as moose and elk, came across the land bridge with the humans; that is, they had already coevolved, made their peace with the conditions of humanity.

An iconoclastic scientist at the University of Arizona first successfully defended the notion that this wave of extinctions was caused by nothing more than the arrival of humans. Before Paul Martin, most scientists believed that climatic upheavals at the end of the last ice age wiped out these mammals. Martin argued they were hunted to death by creatures with nothing more than stone spears and some skill at defeating predators. Martin, whom I interviewed once late in his career, was more or less ridiculed and hounded for promoting this theory; but he persisted, and the evidence mounted, and today his is the generally accepted explanation. If the hunter-gatherer world was Eden, then Eden spread across the world by means of wholesale slaughter. If we are building a model of human nature, this, too, must be factored in.

Furthermore, spread across North America are the archaeological sites of the Clovis people, the name for the culture that first colonized North America. The sites show evidence that this colonization entailed massive waste. People could not store the proceeds of wiping out a herd of mammoths in a single cliff jump, so they ate what they needed and moved on. A harsh new set of conditions prevailed on the planet, and with them came a suite of animals such as bison, elk, musk ox, cape buffalo, tigers, and some camels capable of surviving human tactics. That is, between 40,000 and 10,000 years ago, humans reshaped the composition of earth's megafauna. Think of this as proto-domestication.

For a variety of reasons all tied to humans' special basic abilities, we are unique among predators and foragers. We do not pay nearly as heavy a price for depleting our food sources. We have so many alternatives: we can simply switch food, or move to a new range, or both. Because of this, we are the most ubiquitous species on the planet, inhabiting all of the ecotones for more than 15,000 years. No other animal or plant has done so. This unique ubiquity also explains the trend in the human diet over the long course of our evolution as hunter-gatherers. Biology teaches us that predators will maximize their use of prey with, literally, mathematical precision. Mathematical models have been constructed to analyze how a given animal can best balance the energy costs of obtaining food against the energy obtained. The models work remarkably well, down to the point of predicting how far a lizard will move to snatch a fly. Applied to humans, the model says they should go for the biggest, most docile creatures they can find, and they did. As those became rarer or even extinct, they moved to less efficient prey. We moved toward variety.

Just how far we moved can be read in the well-preserved carcass of an early iron age man buried in a bog in Denmark. His stomach contained the remnants of sixty different species of plants: not the sum total of the range of his diet, merely the range from a day or so, what happened to be ripe or on hand at the moment. Multiply that number through the seasons and across the animal kingdom, and some appreciation for that human's catalogue of sensual clues begins to accrue. Learning to live off hundreds of species of plants and animals required an attention to color, light, shape, and motion that must have bordered on obsession. No wonder we began painting in such fine detail so early in the course of human events. It is as if we were brimming with observation and had to let it all out. The way we preserved our species during our formative years not only made us hunters and gatherers, but painters, singers, and poets, all of the essential sensuality of these arts winding back to food and sex.

In his *Letters to a Young Poet*, Rainer Maria Rilke reminds us of the perils of ignoring this legacy:

> Physical pleasure is a sensual experience no different from pure seeing or the pure sensation with which a fine fruit fills the tongue; it is a great unending experience, which is given us, a knowing of the world, the fullness and glory of all knowing. And not our acceptance of it is bad; the bad thing is that most people misuse and squander this experience and apply it as a stimulant at the tired spots of their lives and as a distraction instead of rallying toward exalted moments. Men have made even eating something else; want on the one hand, superfluity upon the other, have dimmed the distinctness of this need, and all the deep, simple necessities in which life renews itself have become similarly dulled.

A dulling that began in earnest with the invention of agriculture.

WHY AGRICULTURE?

Why agriculture? In retrospect, it seems odd that it has taken archaeologists and paleontologists so long to begin answering this essential question of human history. What we are today—civilized, city-bound, overpopulated, literate, organized, wealthy, poor, diseased, conquered, and conquerors—is all rooted in the domestication of plants and animals. The advent of farming re-formed humanity. In fact, the question "Why agriculture?" is so vital, lies so close to the core of our being that it probably cannot be asked or answered with complete honesty. Better to settle for calming explanations of the sort Stephen Jay Gould calls "just-so stories."

In this case, the core of such stories is the assumption that agriculture was better for us. Its surplus of food allowed the leisure and specialization that made civilization. Its bounty settled, refined, and educated us, freed us from the nasty, mean, brutish, and short existence that was the state of nature, freed us from hunting and gathering. Yet when we think about agriculture, and some people have thought intently about it, the pat story glosses over a fundamental point. This just-so story had to have sprung from the imagination of someone who never hoed a row of corn or rose with the sun for a lifetime of milking cows. Gamboling about plain and forest, hunting and living off the land is fun. Farming is not. That's all one needs to know to begin a rethinking of the issue. The fundamental question was properly phrased by Colin Tudge of the London School of Economics: "The real problem, then, is not to explain why some people were slow to adopt agriculture but why anybody took it up at all, when it is so obviously beastly." Research has supported Tudge's skepticism.

If "why" is complicated, when and where agriculture began is not. We can fairly easily trace each of the modern world's leading crops to its point of domestication—that is, to the natural habitat of its wild ancestors. The great bulk of today's human nutrition, more than two-thirds of it, comes from four crops: corn (or maize, as it is known internationally), wheat, rice, and potatoes. Each can be traced in a clean, crisp line to a spot in the world where both agriculture and a branch of civilization began: maize to central Mexico, wheat to the Middle East, rice to the Yangtze and Yellow River basins of China and the Ganges plain of India, and potatoes to the Andes. Each produced a major, literate, urban civilization: Aztec, Western, Asian, and Inca.

Research has unearthed evidence for broadening this list to include what is now the south-central United States and sub-Saharan Africa. Clearly, domestications and urbanization occurred in both places. Most significantly, rice was domesticated in western Africa, though it was a different genus of the plant that now produces almost all domestic rice. Still, these separate points of domestication followed the patterns of the better-known centers, and it is these common lines that are vital to this discussion, especially as civilizations developed independently and spontaneously in widely separated regions of the planet.

These six places on the globe represent a relatively tiny portion of the lands humans inhabited. So why the leap in these places and only these? The answer has less to do with human range than with plant range. Of all of the species of plants extant, only a tiny subset—a few hundred species—have readily edible parts. These are the candidates for domestication, defined simply as plants with traits that allow them to be easily domesticated and, more important, to be productive enough to justify the effort. Agriculture began in the home range of those plants, because the plants' evolution had already done most of the work. Agriculture was simply opportunism, as research readily demonstrates. For example, Israeli scientists working with wild emmer wheat and barley, the two cereals that are the foundation of Middle Eastern agriculture, found it exceedingly easy to grow the wild varieties in new settings and discovered immediate advantages to doing so, so predisposed were the plants to domestication.

This thin border between wild and domestic then begs the question: What do we mean by domestication? If there was, after all, so little difference between wild and tame, why all the fuss over taming?

The answer often proposed has to do with human intent: agriculture is the result of cultivation, the systematic manipulation of the environment. As scientists began testing this notion, though, they found thousands of years' worth of manipulation by hunter-gatherers preceding agriculture. The most outstanding example is the use of fire, with plenty of evidence that hunter-gatherers worldwide employed massive prairie and forest fires to drive herds of game, remove hiding places, and even stimulate fresh growth that would draw game animals.

Additionally, there is evidence that hunter-gatherers routinely carried seeds with them, and often those seeds would sprout around camps or riverbottoms. We can imagine what that might have looked like from the evidence of historical times. Nomadic Plains Indians, for example, enjoyed about as pure a form of hunter-gatherer culture as has been recorded, living almost exclusively off meat and following herds of bison. Yet they also grew tobacco. Tribes left contingents to spend the growing season stationed in a given river valley, minding the crop.

All of this suggests a class of activity we can call proto-agriculture—practices that look like agriculture but do not cross the line to full-fledged cultivation. There were some clear advantages to this behavior. Detailed work in the American Southwest, for instance, shows that some hunting groups used something very much like agriculture to allow them to occupy key hunting areas during lean times, in this way claiming rights to the territory's meat during flush times. Failure to weather out the bad seasons would have left them no choice but to move on, giving other groups access to the territory. These embryonic forms of farming gave people an alternative, in the same way that the ready availability of Cat Chow makes Tabby songbirds' most lethal suburban predator, not skunks and other wild predators who must move on when pickings get thin.

Colin Tudge argues:

> A creature that can manipulate the environment ever so slightly can hang on in a given location when the more passive creatures are obliged to move out. Over the creature's whole range, this ability to hang on just a little better in a particular location, even when times are bad, means the total population is just a little higher than it would be if the creature was simply a hunter and simply a gatherer of wild plants.

Tudge, in fact, argues that this phenomenon was a key factor in the mass extinction of animals such as the woolly mammoth when hunters first moved into North America. The proto-farmers "could easily, and perhaps gleefully, have pursued the more spectacular creatures to extinction," he writes. I raised this issue in the previous chapter as a question of motive, but this nasty bit of mass human behavior may simply be a question of means that echoes to the present. When we have the ability to kill to extinction, we do so.

Given the deep history of human impact, though, we need to look beyond simple manipulation to define agriculture. Domestication means more than exploitation, which is why there is more than an academic difference between proto-farming and farming. Domestication is human-driven evolution, a fundamental shift in which human selection exerts enough pressure on the wild plant that it is visibly and irreversibly changed, its genes altered. This alteration of plants, and later animals, occurred at each of the key agricultural centers, but probably first in the Middle East, where a site along the Euphrates River shows clearly domesticated einkorn and emmer wheat and barley dating to 9,600 years ago. These wild annual grasses crossed a line—as two others, rice and maize, later would—forever altering their genomes and the terms of life on the planet.

Scientists have a precise definition of that line. The grains of the wild grasses that were the ancestors of these domestic crops had long been in evidence at human camps and settlements in the region. Suddenly the grains became plumper and larger, evidence of selection pressure for larger grains. That is, humans were not merely picking the largest seeds; they had always done that. Instead, the sample changed because there were more larger grains to pick, which meant humans were selecting and planting larger seeds. More telling still was a change in the rachis, the miniature stem that ties each grain to the seed head. Wild grasses have brittle rachises that break easily, which allows the plant to spread its seed and propagate. Gatherers harvest wild seeds by smacking the seed head and catching what falls. Wild rice is still harvested that way in the upper Midwest. To a farmer, though, it is an advantage to have the grain hang on to the seed head; one loses less seed and is able to cut the whole stem and then thresh it by flailing in camp, as farmers have done through the ages. This harvest method selects for seeds that stick to the head, and those seeds cannot easily propagate without human assistance. All three wild grasses from this area along the Euphrates showed this marked change in their rachises all of a sudden, beginning almost ten thousand years ago.

Slightly later, a corresponding change in animals took place, the evidence for which is the bones of butchered beasts. Hunters take a cross-section of sexes and sizes of prey. Pastoralists, however, slaughter young males, keeping the females for brood animals. Accordingly, the farmers' bone pits hold a larger proportion of bones of young males and old females than would otherwise be expected. At the same time, the animals become smaller overall, a result of selection for animals easier for herders to handle. Anyone who has ever dealt with a large, aggressive bull or stallion would understand why a farmer might have a preference for downsizing genes.

So both plants and animals changed, crossing the line of human intervention that defines domestication. Why? Or, to put it another way, what were the necessary preconditions for domestication? One, as I have mentioned, is the presence of candidates for domestication. Agriculture probably happened in the Middle East first because the native wheat and barley were the easiest plants to domesticate. Still, those wild grasses and humans had coexisted in that region for centuries before. What else was going on?

A second precondition of agriculture is catastrophe. By this, I do not mean catastrophe in the human sense; hard times and starvation were not the necessities that mothered the invention of agriculture. In fact, something quite the opposite happened. I mean catastrophe in the biological sense—a natural disaster, or something like one, that resets the biological clock to zero by wiping out an evolved suite of plant life, as happens after volcanoes, floods, and fires. There is no evidence that volcanoes played a significant role, but the creation of agriculture was very much dependent on fire and flood. Agriculture sprouted in the wake of such catastrophes, not just in the Middle East but worldwide. Many domesticated plants are predisposed to grow in flood plains, where periodic inundation provides natural tillage that wipes out competitors. And the slash-and-burn methods (or the more politically correct euphemism "swidden agriculture") on which tropical agriculture has relied into modern times is nothing more than artificially induced catastrophe. The point of the fire is to reset the biological clock.

The earth's elements are in a more or less constant state of upheaval, so that life-forms have had to learn to adapt. Catastrophes happen, and nature has devised strategies to survive. It just so happens that one of those strategies could easily be borrowed by humans. There is a very narrow range of colonizing plants designed by evolution to move in and restart the biological clock after catastrophe. Generally, they are annuals that don't need to survive these harsh conditions year after year. They don't need to persist, to set deep roots. Rather, they invest their resources in building large, easily detached, portable, long-lasting seeds ready to exploit the next sweeping catastrophe. Once these colonizers gain a foothold and provide cover, shade, and organic matter in the soil, a more permanent community of plants dominated by perennials develops. This is the core process of biological succession, the natural maturation of communities. The colonizing annual's strategy of investing its energy in seed doesn't pay off in the mature community, because there is no unoccupied ground in which the seed can grow. The strategy disappears in mature communities. Remember, though, that we call large, easily detached, long-lasting, portable seeds "grain." And as grain is the foundation of civilization, so, by extension, is catastrophe.

A lost, thirsty traveler in North America's West can find water easily enough if he can spot a line of cottonwood trees. They grow among rivers and creeks, prompting the assumption that they are a water-loving species. They are a relatively reliable guide to water, but the assumption is wrong. As industrial society began sending bulldozers to rip roads up western mountainsides, cottonwoods began creeping up those bone-dry roads. It turns out that cottonwoods are not so much water-loving as disturbance-loving.

The same sort of assumption steered thinking about agriculture, because agriculture first flourished along river valleys. And, yes, it did benefit from the water, especially the periodic flooding that layered nutrients in the floodplains, sparking its emergence in the Nile River valley. Just as important to these early efforts, however, was the disturbance created by these floods, nature's plow.

The relationship holds up worldwide. For instance, Bruce Smith, an anthropologist with the Smithsonian Institution, was able to make a compelling case for an emergence of agriculture in what is now the southern United States by examining the changes in seed size and coating in a series of what he calls "floodplain weeds." Wild gourds, sunflowers, and chenopods (a genus of green, leafy weeds of which modern pigweed is an example) flourished along creeks and rivers and were gathered and eventually transplanted to artificially disturbed sites.

Farming's relationship to floodplains was even more pronounced in Asia with the domestication of rice, a form of agriculture that never quite lost its close tie to annual floods. The anthropologist Charles Higham writes, "The accumulated archaeological evidence is unanimous in supporting low-lying aquatic habitat as the most likely location for the transition to rice cultivation."

This close relationship with disturbance, be it fire or flood, meant that agriculture could take its cue and its candidate species from naturally disturbed sites; but to ratchet up to farming would require a significant level of human disturbance. That is, it would require that people congregate in settlements, places where their activity would disturb land, and more important, where people would stay long enough to plant and harvest. Sedentism was a precondition of agriculture. This flies directly in the face of the just-so story that suggests it was the efficiency of agriculture that made settlement possible. In fact, the archaeological evidence suggests quite the reverse: that sedentism—the radical human experiment with staying put—made agriculture possible, and not vice versa.

Sedentism, like flooding, requires a proximity to water. Particular groups of hunter-gatherers became skilled fishermen and settled in stable communities near river mouths. Their dependence on migratory fish such as the salmon was particularly pronounced, then and to the present. Salmon show up in Cro-Magnon paintings—and their skeletons in Cro-Magnon sites—throughout Europe. Cro-Magnon peoples stayed in one place and had enough leisure time to paint, and they painted salmon because salmon were important to them. The rise of art much later among Northwest American Indians is unique among North American hunter-gatherers, suggesting something parallel in the two salmon cultures—a correlation between salmon, sedentism, and art. Fishing a migratory species allows all this. You simply stay put at streamside and the salmon come. Throughout the world, sites along rivers, seas, estuaries, and lakes show layers of shellfish and fish bones below (and thus older than) layers containing evidence of agriculture. These early sedentary people did not have to wander seeking game; currents, the habits of their prey, and the enormous productivity of marine systems like estuaries brought the prey to them.

Agriculture did not arise from need so much as it did from relative abundance. People stayed put, had the leisure to experiment with plants, lived in coastal zones where floods gave them the model of and denizens of disturbance, built up permanent settlements that increasingly created disturbance, and were able to support a higher birthrate because of sedentism.

In the Middle East, this conjunction of forces occurred about ten thousand years ago, an interesting period from another angle. That date, the start of what is called the Neolithic Revolution, also coincides closely with the end of the last glaciation. As I write this, I sit in a spot that was then at the bottom of a huge lake. I live in a valley that held a lake famous to geologists, glacial Lake Missoula. The valley was formed by an ice dam that sat a couple hundred miles from here, and as the glaciers melted, the ice dam broke and re-formed many times, each time draining in a few hours a body of water the size of today's Lake Michigan. That's disturbance. The record of these floods can be clearly read today in giant washes and blowouts throughout the Columbia River basin in Washington State. Within the mouth of the Columbia River, several hundred miles downstream, is a twenty-five-mile-long peninsula made of sand washed downstream in these floods.

When the glaciers retreated, such catastrophic events were happening with increased frequency in floodplains around the world, especially in the Middle East. Juris Zarins of the University of Missouri has suggested that these massive disturbances and floods underlie the central Old Testament myths—the great flood, but also the Garden of Eden. Following a specific description in Genesis of the site of Eden, Zarins traces what he speculates are the four rivers of the Tigris and Euphrates system mentioned there. They would have converged in what is now the Persian Gulf, but during glaciation this would have been dry land. Further, it would have been an enormously productive plain, the sort of place that would have naturally produced an abundance of food without farming.

We call it the Garden of Eden, but it was not a garden; it was not cultivated. In fact, in Genesis, God is vengeful and specific in throwing Adam and Eve out of paradise; his punishment is that they will begin gardening. Says God, "In the sweat of thy face shalt thou eat bread, till thou return unto the ground." God made good on his threat, and the record now shows just how angry he was. The children of Adam and Eve would hoe rows of corn. "To condemn all of humankind to a life of full-time farming, and in particular, arable farming, was a curse indeed," writes Colin Tudge.

At about the same time that the shapes of seeds and of butchered sheep bones were changing, so were the shapes of villages and graves. Grave goods—tools, weapons, food, and comforts—were by then nothing new in the ritual of human burials. There is even some evidence, albeit controversial, that Neanderthals, an extinct branch of the family, buried some of their dead with flowers. Burial ritual was certainly a part of hunter-gatherer life, but the advent of agriculture brought changes.

For instance, one of the world's richest collections of early agricultural settlements lies in the rice wetlands of China's Hupei basin on the upper Yangtze River. The region was home to the Ta-hsi culture that domesticated rice between 5,500 and 6,000 years ago. Excavation of 208 graves there found many empty of anything but the dead, while others were elaborately endowed with goods. The same pattern emerges worldwide, one of the key indicators that, for the first time in human history, some people were more highly regarded than others, that agriculture conferred social status—or, more important, more goods—to a few people.

Some of early agriculture's graves contained headless corpses, corresponding to archaeologists finding skulls in odd places and conditions. Skulls in the Middle East, for instance, were plastered to floors or into special pits. Some of the skulls had been altered to appear older. Archaeologists take this as a sign of ancestor worship, reasoning that because of the permanent occupation of land, it became important to establish a family's claim on the land, and veneration of ancestors was a part of that process. So, too, was a rise in the importance of the family as opposed to the entire tribe, a switch that further evidence bears out.

Coincident with this was a shift in the villages themselves. Small clutches of simple huts gave way to larger collections, but with a qualitative change as well. Some houses became larger than others.

At the same time, storage bins, granaries, began to appear. Cultivated grain, more so than any form of food humans had consumed before, was storable, not just through the year, but from year to year. It is hard to overstate the importance of this simple fact as it would play out through the centuries, later making possible such developments as, for instance, the provisioning of armies. But the immediate effect of storage was to make wealth possible. The big granaries were associated with the big houses and the graves whose headless skeletons were endowed with a full complement of grave goods.

The Museum of Anatolian Civilizations in Ankara, Turkey, holds one of the world's most impressive assemblages of early agricultural remnants, including a reconstruction of a grave from a nearby city once ruled over by King Midas. He was a real guy, and his region was indeed known for its wealth in gold, taken from the Pactolus River. Yet the grave unearthed at Gordium (home of the Gordian knot) once thought to be Midas's (but now identified as that of another in his line) was not full of gold. It was full of storage vessels for grain.

Of course to assert that agriculture's grain made wealth possible is to assert that it also created poverty, a notion that counters the just-so story. The popular contention is that agriculture was an advance, progress that enriched humanity. Whatever the quality of our lives as hunter-gatherers, our numbers had become such that hunger forced this efficiency. Or so the story goes.

We have seen that agriculture in fact arose from abundance. More important, wealth, as distinct from abundance, is one of those dichotomous ideas only understood in the presence of its opposite, poverty. If we are to seek ways in which humans differ from all other species, this dichotomy would head the list. This is not to say that hunter-gatherers did not experience need, hard times, even starvation, just as all other animals do. We would be hard-pressed, however, to find communities of any social animal except modern humans in which an individual in the community has access to fifty, a hundred, a thousand times, or even twice as many resources as another. Yet such communities are the rule among post-agricultural humans.

Some social animals do indeed have hierarchy. Chickens and

wolves have a pecking order, elk a herd bull, and bees a queen. Yet the very fact that we call the reproductive female in a hive of bees the "queen" is an imposition on animals of our ideas of hierarchy. The queen doesn't rule, nor does she have access to forty times more food than she needs; nor does the alpha male wolf. Among elk, the herd bull is the first to starve during a rough winter, because he uses all his energy reserves during the fall rut.

The notion that agriculture created poverty is not an abstraction, but one borne out by the archaeological record. Forget the headless skeletons; they represent the minority, the richest people. A close examination of the many, buried with heads and without grave goods, makes a far more interesting platform for the question of why agriculture. Another approach to this question would be to walk the ancient settlement of Cahokia, just outside of St. Louis, Missouri, and ask: Why all these mounds?

Cahokia was occupied until about six hundred years ago by the corn, squash, and bean culture of what is now the midwestern United States. There are a whole series of towns abandoned for no apparent reason just before the first Europeans arrived. "Mounds" understates the case, especially to those thinking the grand monuments of antiquity are part of the Old World's lineage alone. They are really dirt pyramids, a series of about a hundred, the largest rising close to a hundred feet high and nearly one thousand feet long on a side at its base. The only way to make such an enormous pile of dirt then was to carry it in baskets mounted on the backs of people, day in, day out, for lifetimes.

Much has been made of the creative forces that agriculture unleashed, and this is fair enough. Art, libraries, and literacy, are all agriculture's legacy. But around the world, the first agricultural towns are marked by mounds, pyramids, temples, ziggurats, and great walls, all monuments reaching for the sky, the better to elevate the potentates in command of the construction. In each case, their command was a demonstration of enormous control over a huge force of stoop labor, often organized in one of civilization's favorite institutions: slavery. The monuments are clear indication that, for a lot of people, life did not get better under agriculture, an observation par-

ticularly pronounced in Central America. There, the long steps leading to the pyramids' tops are blood-stained, the elevation having
been used for human sacrifice and the dramatic flinging of the victim
down the long, steep steps.

Aside from its mounds, though, Cahokia is useful for considering
the just-so story of agriculture's emergence because it lies in the
American Midwest, was relatively recent, and was largely contiguous
and contemporaneous with surrounding hunter-gatherer territories.
Like most agricultural societies, the mound builders coexisted with
nomad hunters. Both groups were part of a broad trading network
that brought copper from Michigan's Upper Peninsula to what is now
St. Louis, and seashells from the southeastern Atlantic Coast to
Montana's Sweet Grass Hills. This coexistence gives us a chance to
compare lives by comparing skeletons.

We know from their remains that the farmers were smaller, the
result of general deprivation and abuse. The women, especially, were
smaller. The physiques that make up a modern women's soccer or
basketball team were simply unheard of among agricultural peoples,
from farming's beginnings to only very recent times. On average, we
moderns (and only those of us in the richest parts of the world) are
just beginning to regain the stature that we had as hunter-gatherers,
who throughout time were on average as tall as North Americans are
today.

Part of this decline stems from poor diet, especially for those who
provided the stoop labor. Some of it is inherent in sedentism. Almost
every locale's soil and water are deficient in one mineral or another, a
fact that was not a problem for nomadic hunter-gatherers. By moving
about and taking food from a variety of niches, they balanced one locale's deficiencies against another's excess. This is also true for the
early sedentary cities that relied on seafood. They didn't move, but
the fish did, bringing with them minerals from a wide range of
places.

More important, however, grain's availability as a cheap and easily
stored package of carbohydrates made it the food of the poor. It allowed one to carry baskets of dirt day after day, but its lack of nutritional balance left people malnourished and stunted. The complex

carbohydrates of grains are almost instantly reduced to sugars by digestion, sometimes simply from being chewed. The skeletal record of farming peoples shows this as tooth decay, an ailment nonexistent among contemporary hunter-gatherers.

That same grain, however, could be ground to soft, energy-rich gruels that had been unavailable to previous peoples, one of the more significant changes. The pelvises from female skeletons show evidence of having delivered more children than their counterparts in the wild. The availability of soft foods meant children could be weaned earlier—at one year instead of four. Women then could turn out the masses of children that would grow up to build pyramids and mounds.

They could also grind the grain. Theya Molleson of the Natural History Museum in London has found a common syndrome among these women's skeletons: the toes and knees are bent and arthritic, and the lower back is deformed. She traces this to the saddle quern, a primitive stone rolling-pin mortar and pestle used for grinding grain. These particular deformities mark lives of days spent grinding.

The baseline against which these deformities and rotten teeth are measured is just as clear. For instance, paleopathologists who have studied skeletal remains of hunter-gatherers living in the diverse and productive systems of what is now central California found them "so healthy it is somewhat discouraging to work with them." As many societies turned to agriculture in the early days, they did so only to supplement or stabilize a basic existence of hunting and gathering. Among these people, paleopathologists found few of the difficulties associated with people who are exclusively agricultural.

The marks of agriculture on subsequent groups, however, are unmistakable. In his book *The Day Before America*, William H. MacLeish summarizes the record of a group in the Ohio River valley: "Almost one-fifth of the Fort Ancient settlement dies during weaning. Infants suffer growth arrests indicating that at birth their mothers were undernourished and unable to nurse well. One out of a hundred individuals lives beyond fifty. Teeth rot. Iron deficiency, anemia, is widespread, as is an infection produced by treponemata" (a genus of bacteria that causes yaws and syphilis).

The inclusion here of communicable diseases is significant and consistent with the record worldwide. Sedentary people were often packed into dense, stable villages where diseases could get a foothold, particularly those diseases related to sanitation, like cholera and tuberculosis. Just as important, the early farmers domesticated livestock, which became sources of many of our major infectious diseases, like smallpox, influenza, measles, and the plague.

Summarizing evidence from around the world, researcher Mark Cohen ticks off a list of diseases and conditions evident in skeletal and fecal remains of early farmers but absent among hunter-gatherers. The list includes malnutrition, osteomyelitis and periostitis (bone infections), intestinal parasites, yaws, syphilis, leprosy, tuberculosis, anemia (from poor diet as well as from hookworms), rickets in children, osteomalacia in adults, retarded childhood growth, and short stature among adults.

Such ills were obviously hard on the individual, as were the slavery, poverty, and oppression agriculture seems to have brought with it. And all of this seems to take us further from answering the question: Why agriculture? Remember, though, that this is an evolutionary question.

The question of agriculture can easily get tangled in values, as it should. Farming was the fundamental determinant of the quality (or lack thereof) of human life for the past ten thousand years. It made us, and makes us, what we are. We have long assumed that this fundamental technology was progress, and that progress implies an improvement in the human condition. Yet framing the question this way has no meaning. Biology and evolution don't care.very much about quality of life. What counts is persistence, or, more appropriately, endurance—a better word in that it layers meanings: to endure as a species, we endure some hardships. What counts to biology is a species' success, defined as its members living long enough to reproduce robustly, to be fruitful and multiply. Clearly, farming abetted that process. We learned to grow food in dense, portable packages, so our societies could become dense and portable.

We were not alone in this. Estimates say our species alone uses forty percent of the primary productivity of the planet. That is, of all the solar energy striking the surface, almost half flows through our food chain—almost half to feed a single species among millions extant. That, however, overstates the case, in that a select few plants (wheat, rice, and corn especially) and a select few domestic animals (cattle, chickens, goats, and sheep for the most part, and, as a special case, dogs) are also the beneficiaries of human ubiquity. We and these species are a coalition, and the coalition as a whole plays by the biological rules. Six or so thousand years ago, some wild sheep and goats cut a deal in the Zagros Mountains of what is now Turkey. A few began hanging around the by-then longtime wheat farmers and barley growers of the Middle East. The animals' bodies, their skeletal remains, show this transition much as the human bones do: they are smaller, more diseased, more battered and beaten, but they are more numerous, and that's what counts. By cutting this deal, the animals suffer the abuses of society, but today they are among the most numerous and widespread species on the planet, along with us and our food crops.

Simultaneously, a whole second order of creatures—freeloaders and parasites—were cutting the same deal. Our crowding and our proximity to a few species of domestic animals gave microorganisms the laboratory they needed to develop more virulent, more enduring, and more portable configurations, and they are with us in this way today, also fruitful and multiplied. At the same time, the ecological disturbance that was a precondition of agriculture opened an ever broadening niche, not just for our domestic crops, but for a slew of wild plants that had been relegated to a narrow range. Domesticated cereals, squashes, and chenopods are not the only plants adapted to catastrophes like flood and fire. There is a range of early succession colonizers, a class of life we commonly call weeds. They are an integral part of the coalition and, as we shall see, almost as important as our evolved diseases in allowing the coalition to spread.

In all of this we can see the phenomenon that biologists call coevolution. In the waxing and waning of species that characterizes all of biological time, change does not occur in isolation. Species change to respond to change in other species. Coalitions form. Domestication was such a change. Human selection pressure on crops and animals can be read so clearly in the archaeological record because the archaeological record is a reflection of the genetic record. We reformed the genome of the plants just as surely as (and more significantly than) any of the most Frankensteinian projects of genetic manipulation plotted by today's biotechnologists. The shape of life changed.

Can the same be said of the domesticates' effects on us? Did they reengineer humans? After all, we can see the change in the human body clearly written in the archaeological record. Or at least we can if for a second we allow ourselves to lapse into Lamarckianism. In 1809, Jean-Baptiste Lamarck set forth a pre-Darwinian theory of evolution that suggested that environmentally conditioned changes in an individual would be inherited by the subsequent generations. That is, to put it in modern terms, conditioning changes the genome. This would imply that a spaniel with a docked tail would spawn stub-tailed progeny, or that a weightlifter's children would emerge from the womb with bulging biceps. We know this is false. (Mostly we know—there are some valid neo-Lamarckian arguments.) Many of the changes in humans I've cited above are in fact responses to the changing conditions brought on by agriculture: the malnutrition, disease, and deformed bones were not inherited, but battered into place with each new generation.

These changes are the result of cultural, not biological, evolution. Do not discount such changes as unimportant; culture evolves as surely—and as inexorably and anarchically—as do our bodies, and it does indeed have enormous effect on our quality of life. Poverty is a direct result of cultural evolution, and despite ten thousand years of railing and warning against it, the result is still, as Christ predicted, that the poor are always with us.

By bringing this distinction between biological and cultural evolution into play, I mean to set a higher hurdle for the argument that agriculture was a powerful enough leap in technology to be read in our genome. Agriculture was social evolution, but at the same time it also instigated genuine biological evolution in humans.

Take the example of sickle-cell anemia. As with many inherited diseases, the occurrence of sickle-cell anemia varies by ethnic group, but it is particularly common in those from Africa. The explanation for this was a long time coming, until someone finally figured out that what we regard as a disease is sometimes an adaptation, a result of natural selection. Sickle-cell anemia confers resistance to malaria, which is to say, if one lives in an area infested with malaria, it is an advantage, not a disease; it is an aid to living and reproducing and passing on that gene for the condition. The other piece of this puzzle emerged only very recently. In 2001, Dr. Sarah A. Tishkoff, a population geneticist at the University of Maryland, reported the results of analysis of human DNA and of the gene for sickle-cell anemia. The gene variant common in Africa arose roughly eight thousand years ago, and some four thousand years ago in the case of a second version of the gene common among peoples of the Mediterranean, India, and North Africa. This revelation came as something of a shock for people who thought malaria to be a more ancient disease. Its origins coincide nicely with those of agriculture, which scientists say is no accident. The disturbance—clearing tropical forests first in Africa, and later in those other regions—created precisely the sort of conditions in which mosquitoes thrive. Thus, malaria is an agricultural disease.

There are similar and simpler arguments to be made about lactose intolerance, an inherited condition mostly present among ethnic groups without a long agricultural history. People who had no cows, goats, or horses had no milk in their adult diet. Our bodies had to evolve to produce the enzymes to digest it, a trick passed on in genes. Lactose is a sugar, and leads to a range of diet-related intolerances. The same sort of argument emerges with obesity and sugar diabetes, cardiovascular disease, and even alcoholism. All are widespread in hunter-gatherer groups suddenly switched to an agricultural diet of dense carbohydrates and sugars. The ability of some people to survive these radical foods evolved only slowly through drastic selection pressure.

All of this points to coevolution, which is the deepest answer to the question of why agriculture. The question implies motive, which is to say we chose agriculture because it was somehow better. There are indeed arguments that it was. Yes, life might have gotten harder in the short term, but storable food provided some measure of long-term security, so there was a bargain of sorts. And while the skeletal remains show a harsh life for the masses, the wealthy were clearly better off and had access to resources, luxury, and security far beyond anything a hunter-gatherer ancestor could imagine. Yet we can raise all the counterarguments and suggest they at least balance the plusses, a contention bolstered by modern experience. That is, we have no clear examples of colonized hunter-gatherers who willingly, peacefully converted to farming. Most went as slaves; most were dragged kicking and screaming, or just plain died.

The coevolution argument provokes a clearer answer to the question: Why agriculture? We are speaking of domestication, a special kind of evolution we also call taming. We tamed the plants and animals so they could serve our ends, a sort of biological slavery, but if coevolution is true, the converse must also be true. The plants and animals tamed us. In biological terms, wheat is successful; its success is built on the fact that it tamed humans. Wheat altered us, altered our genome, to use us.

WHY AGRICULTURE SPREAD

A great deal of my daily angst results from leafy spurge, a Eurasian plant most readers have never heard of. I mean this. A single, simple plant tears at my guts. During the growing season, I plot almost daily the next phase of my ten-year war with this species. Most people would call it a weed, yet this is not some simple garden pest; it infests uncultivated, native grassland communities. Biologists call it an "aggressive exotic," a term that fails to communicate to those who are not biologists the seriousness of its threat. Yet when science considers these matters, aggressive exotics generally rank in the top five on experts' lists of threats to habitat. Entire ecosystems are being lost to these plants, a loss that fails to register in the public consciousness, in part because we cannot imagine life swamping life, green killing green.

Leafy spurge made it into the United States during the nineteenth century, during a wave of plant importation from Eurasia. It has been in my region of Montana for about a hundred years, a brief tenure on the landscape, but it has nonetheless achieved a sort of critical mass. Enjoying the principal advantage of exotics—the absence of natural enemies—it multiplies unchecked, gradually supplanting all other plants, including healthy native grasslands. About one-fourth of my seventy acres is at least moderately infested. Using a variety of weapons, including some fairly sophisticated herbicides and an introduced insect that eats the plant's roots, I have been able to fight it to a stalemate, but lately some colonies have shown signs of herbicide resistance, a prospect that chills me. Evolution happens. If that resistance multiplies, the grassland ecosystems I value as habitat for elk and deer will be defenseless.

I've spent a lot of time looking at spurge close-up. Beyond its lack of enemies, the plant has evolved some sophisticated tools for defeating me, even before the herbicide resistance. It relies on rhizomes, fifteen- and twenty-foot-long horizontal roots just under the ground's surface that support a plant twelve to eighteen inches tall. I can pull the plant off a mountainside, but I can't get at the roots in land too steep, rocky, and wild for any sort of machinery. So much energy is stored in that root that it quickly fronts another stem. Worse, when I break a stem, the injury causes the plant to secrete a hormone that promotes growth. Five new stems appear where the previous one broke. These two relatively simple tactics, combined with its unpalatability to most grazers, are allowing spurge to win a war for the land—not just mine, but on rangeland from Nebraska and the Dakotas south through the plains and west through the Rockies.

If a single plant using a few simple tools can overcome modern methods to defeat it, how powerful, then, is a coevolved coalition of exotics?

We have asked why agriculture arose in the first place, but we must also ask why this isolated experiment spread across the planet to the lands of nonagricultural peoples who could see it coming and who, having had plenty of contact with agricultural societies, ought to have known enough to avoid it like the plague it was.

Perhaps this question is best considered against the immense backdrop of the Great Wall of China. According to long-held theory, the Chinese nation conscripted so much forced and slave labor into building the Great Wall in order to protect itself from barbarian hordes—nomads—to the west. To be sure, the Mongols were a problem throughout Chinese history. But some scholars have advanced a different theory: that the wall was built not so much to keep the Mongols out as to keep Chinese peasants in. Certainly anyone who got a good look at equestrian life on the steppe would prefer it to stoop labor in the rice paddies of that intensely hierarchical society.

Agriculture's just-so story, however, says the motion should have been in the opposite direction, that it was contact with farmers that spread the joys of the plow among the heathen. The assumption is that nomads and hunter-gatherers, who usually traded with civilized folk, knew a good thing when they saw it and so simply adopted the farming technology. In other words, a bunch of guys who spent their time running around the woods, hunting and fishing and trading meat for sex, one day saw someone hoeing weeds and said to themselves, "What a fine idea! Let's go do that instead." Is it possible that the technology did not spread entirely by adoption, that hunter-gatherers were wiped out or displaced by an advancing agricultural imperialism? The record suggests that although some adoption did occur, by and large farming spread by genocide. Those hunter-gatherers who apparently chose to adopt it tended to pick and choose, to assimilate only those parts of it that were attractive to them— pastoralism most often. It is easy to understand how a shift to pastoralism, herding domestic animals, was a relatively small and natural step for hunter-gatherers, who already followed wild herds across their native ranges. As game became scarce, some hunter-gatherers got closer and closer to wild herds and then began to integrate them into their wanderings, so that some of the wild goats, sheep, and cattle became domestic. Their association with the animals amounted to selection pressure that fostered domestication, but the domestication of wandering herds was not a transitional step to agriculture, to growing crops. Domestication of crops came first, an order that holds up in all of the major agricultural centers. Row-crop farmers, not hunter-gatherers, were the first to domesticate livestock. Pastoralism as a separate activity later spread from the agriculturalists to the nomads, a case of selective adoption of an agricultural trick. Choosing what worked in their milieu, especially in arid grasslands, nomads reached for the shepherd's crook but ignored the plow. By and large, though, when agriculture spread as a full-blown system of technologies—plows, wheat, cattle, cities, and priests—it did not diffuse among people but displaced them, and "displaced" is a euphemism.

To some degree, all of the major systems of agriculture did spread. By the time of the Spanish conquest, maize had made its way from its site of domestication north to the upper Mississippi and Missouri basins and south to the Andes. That long north-south axis, however, placed some pretty severe constraints on the spread of the full complement of domesticates in the New World. Not all of Mexico's crops could travel across the tropics and the equator to Peru and vice versa. But some, such as maize, could.

The rice culture of Asia did its share of moving as well, largely south into Southeast Asia, but also west as far as North Africa. Rice was in the Middle East three thousand years ago. Nonetheless, rice farming is a lowland activity that works best in major river valleys and estuaries of the tropics. There were constraints with rice, but not nearly so many with the special coalition of crops and animals developed in the Middle East: wheat, barley, goats, sheep, and cattle were all far more portable. In addition,their domestication occurred at the very southern edge of a vast, fertile plain that stretched from the edge of Siberia, across what is now Russia, west through northern Europe and to the Atlantic. This, of course, is an east-west axis spanning a broad stretch of a temperate zone, with similar conditions every bit of the way. The coalition led by wheat and cattle didn't have barriers to jump, at least not in Eurasia; it found fertile ground all the way, the argument at the center of Jared Diamond's *Guns, Germs, and Steel* and in Alfred W. Crosby's earlier book, *Ecological Imperialism: The Biological Expansion of Europe, 900–1900*.

This culture began by spreading from what is now southern Turkey and northern Iraq north to the area between the Black and Caspian seas, near the Caucasus Mountains. It refined its techniques here, and formed its own language group, Indo-European, the mother tongue that would found Sanskrit in India and English at the opposite edge of the axis. Another name for the wheat-beef culture is "Caucasian."

Much of this early spread encountered no resistance. It colonized lands that were vacant or only sparsely settled by hunter-gatherers, who moved on. This was true, for instance, in Greece and Italy, which lacked the sort of hunting conditions and megafauna that would have supported large groups of hunters. Here, diffusion was simply a matter of agriculture's increased birthrate sending colonists into unoccupied lands. About the same time, there is evidence of some pastoralism spreading among hunter-gatherers, specifically the Cro-Magnon people near the Mediterranean. To the degree that this happened, it was a big change for the people who had occupied that region for at least thirty thousand years before the farmers moved in. These Cro-Magnons are the people responsible for the famous cave art of France, and their art itself tells us they probably had very little incentive to change during all those years. Ian Tattersall, curator of the department of anthropology at the American Museum of Natural History, summarizes their existence in *Becoming Human*:

> The late Ice Age inhabitants of southern France and northern Spain lived in an area greatly favored by geography. This relatively sheltered limestone region boasted a huge range of habitats, from hilly crags, where ibex abounded, right down to the valley floors, where rivers teamed seasonally with migrating salmon. Perhaps never before had fully modern humans ever lived in such a productive and varied environment; and if so, the visual stimuli to artistic production had never been so great.

This relatively peaceful spread of agriculture accelerated and became more aggressive at the opposite edge of Europe, in what is now Hungary, about seven thousand years ago. The original Caucasian farmers adapted very quickly to the rich loess soils of the Hungarian plain (which are similar to Iowa's rich loess soils), and there finetooled a system based in beef and wheat. These people had a taste for the virgin soils of a frontier occupied by hunting and gathering natives, and conquest here would be swift and violent, a foreshadowing of the developments in the American Midwest in the nineteenth century.

The archaeological term for these neolithic cowboys is Linearbandkeramik, derived from the German word for their distinctive pottery and commonly abbreviated as LBK. They were Europe's original wheat farmers, bringing themselves and their bit of foodstorage technology in from what is now southern Turkey. The most revealing marker of their culture's spread was this pottery's sudden ubiquity across Europe. Sites dated from a relatively brief period of 6,700 years ago to 5,900 years ago occur throughout the northern European loess plain, from France to the Ukraine and south into the Danube basin. The anthropologists T. Douglas Price, Anne Birgitte Gebauer, and Lawrence H. Keeley write, "Perhaps the most striking feature of LBK is its remarkable homogeneity in material culture, settlement pattern, and economy over this huge region. . . . Such remarkable homogeneity must evidence either a very rapid spread or a pathological conventionality or both."

It was as if future archaeologists were to find dumps from Des Moines full of Tupperware.

The same group of anthropologists concluded that this culture's sweep through Europe took no more than three hundred years, a blitzkrieg by the standards of the day. And it is appropriate to employ the war metaphor here, in that the record suggests, contrary to conventional ideas about rational and peaceful cultural diffusion, that there was almost no intermixing among the wheat farmers and the salmon-eating, cave-painting Cro-Magnon already resident.

The curious part of this is that there was probably not an inherent ecological reason for conflict. That is, the LBK people didn't blanket the region, at least not at first, but tended to cluster in villages where loess soils were concentrated, leaving the river-valley bottoms and mountains untouched. That would have left a viable niche for hunter-gatherers. A coexistence with mutually beneficial trade could have developed between the two cultures, but the record says it didn't. There is almost no record of Cro-Magnon artifacts in LBK villages and vice versa. Cro-Magnon sites seem to cease being occupied at about the time of LBK arrival. In fact, the record seems to show that the Cro-Magnons maintained a sort of buffer zone between themselves and the newcomers, leaving even in advance of the advancing farmers.

The exception to the absence of artifacts from one culture in settlements of the other is evidence that the two sides swapped spear points, probably not as trade goods. "All these artifacts are weapons," note Price, Gebauer, and Keeley, "and there is no reason to believe that they were exchanged in a nonviolent manner. . . . The evidence from the western extension of the LBK leaves little room for any other conclusion but that the LBK-Mesolithic interactions were at best chilly and at worst hostile."

Can we then conclude that the farmers killed all the Cro-Magnons? There are a couple of ways to pursue this question. The record of human migrations is well-preserved in genes and in languages, both of which evolve along similar principles. The Italian scientist Luigi Luca Cavalli-Sforza, now a professor of genetics at Stanford University, has spent a lifetime chasing these questions across the disciplines of archaeology, anthropology, genetics, molecular biology, and linguistics. He leans heavily on the latter three when he summarizes his findings in his 1995 book *The Great Human Diasporas*. The genetic record matches the archaeological one, he found, and is consistent with the notion that the farming culture originated in the Middle East, from which it spread very quickly across Europe. These people's genes dominate in modern Europe. There was some genetic mixing, but this does not mean the spread was amicable. The spread of these same farmers, my ancestors, in North America was genocidal and violent, yet my seemingly Caucasian body carries the genes of my Chippewa Indian ancestors. Conquering people take brides, slaves, and concubines from the conquered.

The intriguing part of Cavalli-Sforza's genetic analysis, though, shows a small island of genetic resistance, a surviving pocket of what he believes are Cro-Magnon genes, in southern France and northern Spain, the area now or recently occupied by Basque-speaking people. The record says those same people once occupied the broader area of southern France that holds the famous cave paintings. The linguistic case is even more interesting. Almost all of Europe speaks a derivative of the Indo-European language that began near the Caucasus Mountains and hybridized later in the wake of both Greek and Roman expansions. There are, broadly speaking, only two exceptions. One is the Uralic languages: Finnish, Estonian, and Hungarian. The other is Basque, a linguistic island. Linguists have dug deeply for languages related to Basque. The closest they have found are ancient tongues, including Na-Dene, spoken by some North American natives, and another spoken by some Sino-Tibetan peoples. Writes

Cavalli-Sforza, "Basque descends from the language spoken by the first modern humans to come to Europe, the Cro-Magnon." It may be that the plight of modern Basques—their long, bitter struggle with the ethnic majorities in Spain—stretches back to spear points in LBK camps.

About 5,500 years ago, houses in southern Scandinavia got bigger. Storage areas were attached to individual, not communal, homes. Burials became more elaborate. One would expect to find wheat and barley on the scene, and they are. Agriculture arrived, but not in the same manner as it had on the plains to the south. In the areas that are now Denmark and southern Sweden, it appears that the existing people were not overwhelmed and conquered by advancing farmers, but slowly adopted agriculture over the course of about a thousand years, as opposed to three hundred in the much vaster European plain.

What was different? Probably the answer is sedentism. The people in Scandinavia already lived a settled life before farming came. They were among those people throughout the earth who had gathered at the edge of productive rivers, estuaries, and seas. The archaeological record indicates they lived rather abundantly on fish, mollusks, and marine mammals. In fact, their settlements looked a lot like those that had given rise to agriculture much earlier in the Middle East. Their stable economy seems to have given them some firm basis on which to resist the onslaught of the farmers. In addition, they lived at the edge of what was at the time a dense forest, becoming denser as sea levels rose with the decline of glaciation. This was not the open loess plain that allowed agriculture to spread unchecked. In Scandinavia, the wheat-beef complex met its northern ecological limits. It did slowly take root in this land, but only as the people who already lived there and knew best how to survive there retooled it to fit their needs.

There is clear evidence of trade in this region at the beginning of this period. Artifacts made by and imported from the farmers to the south, including polished stone axes, amber, copper jewelry, and copper axes, appear in Scandinavian settlements. This suggests that there was plenty of contact between the two groups, enough to allow traders to occasionally borrow technology and domesticated animals. At the same time, there is evidence that food from the sea was becoming scarcer, so the Scandinavians would need some of the food agriculture could provide. That's not to say all of this happened in peace and harmony. Cemeteries are full of murder victims, both before and after farming came, and the period was likely as violent as any other. But while there is evidence of cannibalism among both groups, the northern transition to agriculture was not prosecuted by conquest and genocide. The violence was retail, not wholesale.

The European expansion of farming thus provides two models for looking at what would come after. The parallel records in the United States lie in the remaining Indian reservations on the vast plains near my home in Montana. They are poverty-wracked, long-term concentration camps for hunter-gatherers whose bison, elk, and deer were largely wiped out by advancing wheat farmers. By contrast, the natives along the northern Pacific Coast of British Columbia and Alaska, remnants of a salmon culture that historically derived some 95 percent of its nutrients from the sea, live on in a traditional life that is relatively intact. These are not concentration camps but communities with a viable economy. Art forms that developed before white settlement persist, as does a reliance on salmon fishing. The thick coastal forests of the Cascades won't grow wheat. In places like Prince Rupert, British Columbia, both natives and whites still do what people have done there for five thousand years: they fish. A city council member is as likely to be Tshimshian as Swede. (Many of the immigrant communities here are Scandinavian.)

In fact, the similarities to Scandinavia's transition to agriculture are strong enough to suggest that they mark not a parallel but an extension of the same expansion. After all, the conquerors of North America are descendants of the very same people who swept the Cro-Magnons out of Europe.

"Perhaps European humans have triumphed because of their superiority in arms, organization, and fanaticism, but what in heaven's name is the reason that the sun never sets on the empire of the dandelion?" This is one of the fundamental questions asked by Alfred W. Crosby in *Ecological Imperialism*. Crosby traces the extension of the expansion that began five thousand years before in Europe proper to create what he calls the "neo-Europes": the temperate regions of North and South America, New Zealand, and Australia. (Some would include South Africa as a neo-Europe as well.) The core of the neo-Europes is made up today of New Zealand and Australia plus the United States, Canada, Argentina, southern Brazil, and Uruguay. These are the nations now dominated, if not demographically, certainly economically and politically, by Caucasians, as demonstrated most obviously by the prevalence in these places of European surnames. Less obvious is an even more important marker of dominance. In 1998, the seven countries on the list accounted for about two-thirds of all wheat exports in the world, and about 70 percent of all maize. These are the most important two of the big three grain crops that account for two-thirds of the world's nutrition. That is to say, the new Europes drive the world's agriculture. The dominance does not stop with grain. These seven countries, plus the mother ship—Europe—accounted for three-fourths of all agricultural exports of all crops in the world in 1999.

The single outsider that can claim some standing in this group is China, which is becoming a major player, not in exports of wheat and maize but, more important, in their production. Although it is not a neo-Europe, China feeds itself not just with rice but as the world's leading producer of wheat and a significant grower of maize. China has vast stretches of temperate lands at the same latitude as Europe's wheatfields. Its strong, long history of rice agriculture allowed it to successfully resist European invasion, but it has taken advantage of its ecology and long-standing trade with the West to prosper with European crops.

I lay all of this out to justify what may appear to be a Western bias, or Eurocentrism, in my narrative. Certainly there is more to agriculture, and to culture itself, than what arose in Caucasian Europe. The modern story of imperialism and conquest, however, is decidedly European. We can track that dominance through the spread of wheat and corn, but it is just as valid to do so by following the spread of the English and Spanish languages, of wheat-, milk-, and sugar-tolerant digestive systems, of smallpox-tolerant immune systems, of horses, leafy spurge, beefsteak, or, as Crosby does, dandelions:

> By 3,000 years ago, give or take a millennium or so, "super-man," the human of the Old World civilization, had appeared on earth. He was not a figure with bulging muscles, nor neces-sarily with bulging forehead. He knew how to raise surpluses of food and fiber; he knew how to tame and exploit several species of animals; he knew how to use the wheel to spin out thread or make a pot or move cumbersome weights; his fields were plagued with thistles and his granaries with rodents; he had sinuses that throbbed in wet weather, a recurring problem with dysentery, an enervating burden of worms, and impres-sive assortment of genetic and acquired adaptations to dis-eases anciently endemic to Old World civilizations, and an immune system of such experience and sophistication as to make him a template for all humans who would be tempted or obliged to follow the path he pioneered some 8,000 to 10,000 years ago.

Farmers in this period may not have lived by "bulging forehead" alone, as Crosby indicates, a characterization that challenges the technology-centered explanations for agriculture's spread that have dominated our thinking. Still, technology played an enormous role, as did the east-west axis of agriculture emphasized by others. By two thousand years ago, agriculture had become a continuously linked band stretching all the way from Spain and the British Isles in the West, through China, Japan, and Indonesia in the East, including all of the Indian peninsula. This trade region included a narrow band of North Africa along the southern Mediterranean that was thoroughly integrated into the whole. Imperial Rome, for instance, got most of its grain from Egypt. No longer, then, was there a distinct European agriculture. The main mass had become truly Eurasian, bordered by a band through what is now Ethiopia in eastern Africa and a sub-

Saharan strip of purely African farming. Separate still, but already thriving, were agricultural societies in the northern half of South America, all of Central America, and, in North America, smaller regions of the American Southwest and the corn culture of what is today the southeastern United States. Finally, there was a thriving agricultural center in the West African region known as Senegambia. The people there had independently domesticated a species of rice separate from Asian rice. All of this combined to support what became, by five hundred years ago, the world's first half-billion people.

A good part of what powered this expansion was technological leaps, aided and abetted by widespread trade and contact across the main Eurasian band. The Chinese, for instance, had adopted Europe's big invention—wheat—by three thousand years ago, but they also contributed greatly to the technology used by European farmers during the same period. Arguably the biggest technological leap of the era was the invention of the horse collar about 1,500 years ago in China. Before this, tillage in both Europe and Asia had depended heavily on oxen and a throat-and-girth yoke that suited those ponderous beasts. The same harness was used on horses, but was so inefficient that it greatly limited the load and mobility of these much faster animals. A smaller horse collar allowed a quantum leap in the load a horse could pull, so fields became larger and more widespread almost immediately. This invention traveled quickly from China to Europe.

Agricultural historian L. T. Evans writes:

> The introduction of the padded horse collar also had profound effects on agriculture, by allowing the greater strength, endurance, and mobility of horses to be utilized. Cultivation became more timely. The farmers could live further from their fields and therefore in larger villages with more diverse opportunities, while the sale and faster transport of surplus produce encouraged commerce and the growth of cities.

In short, this simple innovation probably had as profound a role in reshaping the world of that time as internal combustion and tractors have had in our own.

It also made the two legs of European agriculture into a tripod. Wheat, cattle, and horses formed the trinity that, when carried by humans to distant fertile ground, would eventually colonize the world. These three, however, did not act alone. In fact, the broad region of Eurasian farming developed a coalition of domesticates and fellow travelers that would be every bit as important in this conquest as the three species that spearheaded it. Just as in the case of the horse collar, a key to this evolution was the broad east-west band that allowed technology and new or newly vitalized species to spread. The wide geographic band allowed for diffusion, evolution, and synergy. This was certainly true of technologies like cotton spinning (which began in India but quickly spread to the Middle East) but also of new crops and varieties of crops that developed to match slight climate variations.

While the Chinese were adopting rice, for instance, the people of the Indian subcontinent were taking what were to become their mainstream crops of sorghum and millet from Africa, cucumbers from the Himalayas, cannabis from Central Asia, and sugarcane, bananas, and taro from Southeast Asia. Humanity, even in prehistoric times, was already deeply involved in an African-Eurasian economy (if not a global economy). The trade that made this possible also allowed for a variety of freeloaders and stowaways to work their way through the system. The cultivation of storable food created an enormous new niche for whole classes of parasites, insects, birds, rodents, and vermin that fed on those growing fields and stored crops. The best example is the common rat, a European contribution that would become a strong ally of agriculture in subsequent conquest.

A new niche opened for a class of plants as well. Agriculture is based on those very few plants that thrive on disturbance and produce lots of seed. There is a wider group, however, that thrives under catastrophic conditions but without the payback of seed for humans, a class of plants commonly called weeds. Once catastrophic agriculture cleared the path for them, weed species once relegated to floodplains or the edges of forest fires could run in that same contiguous band as farming did, from Britain to Japan. Within that territory, they could evolve, develop, combine, and invent in tandem with humans and human technology. Weeds, like humans, traveled, traded information (genes) with other weeds, and got better at what they do.

Crops, weeds, vermin, and pests developed into a coalition, a cast of characters that depended on one another and traveled together, led by the plow; however, by no means was the most formidable element of this phalanx its visible players.

Our consciousness of the devastation smallpox wrought has focused on the New World and the inherent susceptibility of Native Americans, but the disease is far from a New World problem alone. It also devastated some remote corners of Eurasia that agricultural society reached late in its period of expansion. Smallpox first appeared among the Ostyak, Tungus, Yakut, and Samoyed peoples of Siberia in about 1630, plenty late enough for us to have a clear record of its effects. Crosby writes that it swept "like a scythe through standing grain. The death rate in a single epidemic could soar past 50 percent. When it first struck Kamchatka in 1768–69, it killed two-thirds to three-fourths of the indigenes."

The pattern repeats itself with a series of infectious diseases three centuries later. In 1943, the Alaska Highway pushed into parts of the north that had previously had no significant contact with Europeans. There followed a year-long epidemic of measles, German measles, dysentery, jaundice, whooping cough, mumps, tonsillitis, and meningitis, Crosby reports. A similar event among Eskimos and Indians spread throughout northern Quebec in 1952, killing 7 percent of the people, a relatively low death rate, probably thanks to modern medicine. Modern medicine, however, also gave us better diagnoses, and therefore a better catalogue of the maladies that played a major role in the conquest of the globe by the agricultural Europeans.

Scars on bones tell us that agricultural people themselves dealt with disease from the beginning of agriculture. So does the written record; epidemic diseases were probably at the root of the biblical plagues in Egypt. The first book of Samuel in the Old Testament speaks of a plague, probably bubonic plague. Throughout Eurasia, farming people dealt with epidemics of smallpox, influenza, malaria, measles, chicken pox, and other devastating illnesses long enough to develop immunities. These immunities were a key factor in the domestication of humans. Infectious diseases were selection pressure, and a lot of farming Eurasians no doubt went the way of the aborigines in the neo-Europes. Those who survived, however, were shielded. So the just-so story of European expansion that depicts a rapid march through the primitive world thanks to a superior technology, an evolved and more efficient way of life—seen in North America as a "manifest destiny"—must not credit only technology-driven weaponry and religious zealotry; it must be revised to take into account also the preceding conquest wrought by disease. We are beginning to understand that by the time the conquistadors struck the Andes or Custer reached the Black Hills of South Dakota, only shadow populations of natives remained. The Indian wars got the headlines, but they were mopping-up operations. The shock troops were diseases, especially smallpox, aided by weeds and a few other members of catastrophic agriculture's evolved coalition. Crosby and Diamond both make convincing cases that, at its roots, conquest was biological, not technical.

All of these diseases moved through these untouched territories according to the rules of what are called "virgin soil epidemics." Under those rules, nearly every person who falls sick dies. Infection rates are death rates, partly helped by the fact that infection is so widespread. So many people are sick that few healthy people are left to attend to the sick and to help them survive. Crops go unattended; no one hunts; food sources dry up.

When whites arrived, the indigenous people of the New World had already established lengthy trading routes, and the vanguard of Europe spread along those routes in advance of the colonizers. Further, tribes and villages hit with smallpox or some other epidemic would often take the nomads' solution to adversity and move, and this movement became yet another ally of the pathogens.

The vanguard of the United States' conquest of its territory was the Lewis and Clark Expedition, which reached the mouth of the Columbia River in the winter of 1805–6. Lewis and Clark were there to detail what we now regard as initial conditions among the natives, but Captain James Cook had been there almost thirty years before, the first white contact in the Pacific Northwest. He saw towns that were depopulated even then, and he specifically reported seeing natives bearing smallpox scars. Lewis and Clark spent their first winter on the high plains among the Mandan. The early chapters of their journals detail the tribe's thriving existence. Smallpox later wiped out the Mandan.

The conquistador Hernando de Soto swept through what is now the southeastern United States between 1539 and 1542 to find what we can only describe as an advanced agricultural society, only slightly less developed than Aztec Mexico at about the same time, and with a similar foundation for its agriculture. Based on his reports, the population of Florida was estimated at the time to be nine hundred thousand. He found cities, trade, and royalty along the (later-named) Gulf Coast of Florida between Tampa Bay and Mobile Bay, along the Georgia Coast, along the Mississippi River, in eastern Mississippi, and in southern Arkansas and Louisiana. Then he left. When the French arrived in the same area in the eighteenth century, they found a few primitive villages. The fields of maize were gone, and bison, which de Soto had not seen, had moved back into the deserted landscape. De Soto left without conquering the region, but he left smallpox behind, and disease wiped the slate clean of natives to allow subsequent colonization.

Crosby records a sort of reverse example of the spread of smallpox that helps explain its virulence for natives, and in the process credits the disease with undermining the Anglican Church in America. Unlike other denominations, the Anglicans (fittingly) maintained a rule that would-be bishops must be trained in Britain. Thus, aspiring second- and third-generation colonists would make the trip back to the motherland. Smallpox, however, was not widespread in white settlements in the New World. Immunity to smallpox is not strictly hereditary but accrues in childhood, when children are better able to fend off its effects and when they tend to contract the less virulent forms that confer immunity to the more lethal forms. When the would-be bishops hit the streets of England as adults, they usually caught smallpox and died.

Dead clergy notwithstanding, at least some New World settlers considered smallpox a net gain for their god. John Winthrop, the first governor of Massachusetts Bay Colony, was blunt about the matter: "For the natives, they are neere all dead of small Poxe, so as the Lord hathe cleared our title to what we possess."

If that long Eurasian east-west axis was such an advantage to the supercontinent, then, presumably, an even longer axis would increase the effects that much more. The larger axis is not simply a thought experiment, but existed periodically during glaciation, when sea levels dropped to make the Bering Strait a land mass, Beringia. On this planet, glaciation is not the exception but the norm. Over the last billion years or so, a regular cycle has governed the earth, characterized by glaciations lasting sixty to ninety thousand years, interrupted by brief interglacials of roughly ten thousand years. All of the history of human civilization has been compiled within the present interglacial, which, at the moment, is about out of time.

We know that the last glaciation brought humans to the New World. We have already seen the upheaval this brought for the large mammals native to the continent. During this same period, however, a second bit of commerce occurred that was to set up the expansion of agriculture in Eurasia and its spread in the New World more than ten thousand years later. Eurasia got horses from North America. Horses evolved in the New World as part of the rich complement of fauna of the great interior grasslands. Paleontologists tell us that the development of flat grinding teeth in horse fossils about sixty million years ago signals the rise of the Rocky Mountains. That cordillera formed a "rain shadow" that dried the flat interior, making for arid grasslands. Grass is tough and requires flat, grinding teeth. So the Rockies "built" a creature ideal for colonizing the similar steppes of Central Asia, a development the kin of Genghis Khan would later celebrate.

Horses were among the doomed megafauna of the New World, hunted to extinction by the human immigrants moving across Beringia. The species, however, had already migrated to Eurasia dur-

ing one of the ice ages and eventually became the premier domesticate of Eurasia, greatly enhancing farming and transportation (on
which agriculture's cities came to depend) as well as warfare and
conquest, equally vital institutions of agricultural society.

The power of the horse to reshape society was immediately
manifest in the New World when they were repatriated there by
Europeans. Like smallpox, the horses introduced by the Spanish
conquistadors spread in advance of the whites themselves. This was
especially true in the interior grasslands, where horses had evolved
originally. So fitted are these animals to this environment that descendants of those introduced Spanish ponies still survive as feral
herds in the Great Basin. The natives found them exceedingly useful
for hunting, so much so that once they were introduced, long-settled
corn-farming communities of the Mississippi valley gave up farming
to hunt bison on horses. It is tempting to record this as a footnote of
agricultural history suggesting that, given a choice, people would
sooner hunt than hoe, but something else is probably at work here.
As smallpox spread in these communities and among existing nomads, it probably became difficult to maintain the social organization
that farming required, just as it became an advantage to be on the
move. In addition, hunting methods before the horse required mass
drives on foot, which also depended on a certain amount of social organization and structure. A lone, mounted hunter, however, could be
effective in hunting bison, particularly after rifles arrived.

Interestingly, the experience in the North American plains runs
counter to the spread of European agriculture in the rest of the temperate regions of the globe. The difference was bison. There were
close relatives of and species of bison around the world, just as there
were ancestors of bison in the fossil record in North America. The
closest relative extant is the European steppe bison, now limited to a
relict population in Poland. Thus, there is some cause to believe that
the particular species that is the American bison evolved with human
hunters in Eurasia and crossed the land bridge to replace the ancient
species hunted to extinction by the first immigrants. If this is the
case, then the American bison is, in the long view, an exotic. In any
event, North America was unique. Humans hunted the big grazers to

extinction when they first arrived in North America, New Zealand, Australia, and South America, leaving an enormous, unfilled niche, but only in North America did the immigrants bring a replacement, the bison, to fill that void. The presence of bison allowed the North American nomadic hunters to maintain a strong economy based in bison and to resist white encroachment until the very end of the nineteenth century. This economy collapsed only after American colonizers developed a deliberate policy of exterminating the bison, reducing its numbers from about seventy million before white settlement to fewer than fifty in a half century, all in small private herds or in Yellowstone National Park.

Elsewhere, horses and other big grazers stepped into an empty niche and almost unlimited possibilities for expansion. In South America, the biggest native grazers were llamas and alpacas—camels really, but little and docile ones, hardly competition for horses. The Spaniards attempted to settle in what is now Argentina in the early 1500s, but failed. They did, however, leave behind some horses. When they returned to try colonization again in the 1580s, they found horses in abundance. A traveler at the turn of the seventeenth century reported horses "in such numbers that they cover the face of the earth and when they cross the road, it is necessary for travelers to wait and let them pass, for a whole day or more, so as not to let them carry the tame stock with them." In 1744, a Jesuit in the pampas reported herds of feral horses so numerous that it would take three hours for them to pass by "at full speed."

On North America's East Coast (where bison were scarce), feral horses reached such numbers that they were a nuisance—but a convenient nuisance. Frontiersmen had cheap mounts for the taking, simply by capturing wild animals. Horse colonization followed a similar pattern in Australia's great grasslands. By the mid-nineteenth century, wild horses had become so abundant as to be declared "a weed among animals." Australians killed them for their hides, which fetched only about four shillings apiece. Once the hide market became glutted, the animals had almost no value and settlers knifed or otherwise wounded them so they would run off and die elsewhere.

At the same time, horses were enormously expensive in Europe, their possession out of reach of the average farmer. Because horses were the very engine of the agrarian economy, their availability in the New World was key to the economic potential of the region. Crosby sums this effect: "Horses in such profusion, tame or feral, existed nowhere else on earth. Their abundance shaped the societies of the pampas more firmly and more permanently than the discovery of gold would have."

The unoccupied niche for big grazers was as welcoming to two of catastrophic agriculture's most crucial and long-standing allies, cattle and sheep. They spread as the horses did, in a profusion analogous to the virgin-soil epidemics that spread the key pathogens of the time.

Cattle did not immediately adapt to the tropics, where they were first introduced by Columbus in 1493, but they survived. Subsequent introductions by conquistadors in South and Central America and the southwestern United States in the sixteenth century, however, brought them to temperate grasslands more to their liking. They literally went wild, their populations doubling every fifteen years. So well were they adapted that they could propagate without human aid, so that when, for instance, Jesuits abandoned a mission on the pampas in 1638, the five thousand cattle left behind didn't skip a beat. By 1700, there were an estimated forty-eight million feral cattle on the pampas, roughly equivalent to the number of bison on the Great Plains.

The animals did nearly as well in the eastern United States, with feral herds numbering in the thousands. Many were captured and slaughtered, rendering some settlers wealthy beyond the imagination of their European forebears. At the same time, though, the cattle were independent enough to proceed with settlement on their own, in advance of their human colleagues. They wiped out native grasslands, in many cases before whites ever saw them, paving the way for the European flora to follow. A horde of cattle hooves works like a plow.

Essentially the same story played out in Australia in the eighteenth and nineteenth centuries. By the end of the period, feral cattle would number in the millions in the interior grasslands. The only thing that prevented Australia from being overrun with cattle was sheep, which became increasingly important in the subcontinent as well as in the rest of the neo-Europes as textile manufacture mechanized in the nineteenth century. This development made grasslands economically important enough to wrest from cattle and cattlemen.

While the grassland troops were the grazers, swine worked around the edges, preferring more tropical and wooded areas. Pigs found the living easy in Brazil, the West Indies, Central America, the southern United States, and Australia. Just a decade after their introduction in Hispaniola by Columbus, their prodigious reproductive capacities caused feral hogs to be listed as "infinitos." They literally swarmed the hillsides, vacuuming them of vegetation.

From a human point of view, the mammals were the most significant introductions of the period, but moving with them was a tide of fauna and flora with probably as much significance for native species. From the beginning, explorers brought honeybees to the New World, the bees spreading in advance of human settlement just as the mammals did. Clearly, this paved the way for European crops and weeds, many of which are pollinated by honeybees. But these introductions had profound implications for native plants as well. Pollination is a finely wrought relationship, providing some of the most intricately entwined examples of coevolution. There are, for instance, trumpet-shaped flowers of a certain color that attracts a particular species of hummingbird, which, in turn, has a beak of exactly the correct shape and size to pollinate that specific flower. The New World obviously had its own pollinators and plants dependent on them, but as honeybees spread, they broke up some long-standing relationships, likely wiping out vulnerable species even before Europeans knew they existed. Some native species, like eucalyptus in Australia, did benefit from honeybees' efforts, but even in such cases the ecological balance was altered, allowing certain native species to outcompete others.

One need not know very many plants to appreciate their impressive power to tell the history of a place. One can, for instance, travel to a trail-top summit in Montana's Glacier National Park, one of the most pristine and protected places in the United States, and find among the native flora the distinctive blue-tipped leaf and airy seed head of Kentucky bluegrass, the same plant settlers found growing in what would be Kentucky when they first crossed the Appalachian Mountains. But Kentucky bluegrass is not from Kentucky; it's from Europe and was introduced on the East Coast when marauding cattle made a place for it. Like smallpox, bees, and cattle, it traveled ahead on its own. It grows on summits in Glacier Park because the area was used by horse packers, and they deliberately brought the seed and scattered it so that their horses might have something to eat.

The practice of seeding to promote conquest began with conquest itself. Long before Columbus made his discovery, Spain began practicing for colonization with the conquest of the Canary Islands, which were immediately "seeded" with sheep and swine. Throughout the period of exploration, mariners typically carried these animals aboard and dropped them on strategic islands along the way, along with handfuls of grasses and other plants they thought might take root and provide sustenance so that sailors would have a food supply when they returned years later.

The last of the neo-Europes to be settled was New Zealand, and by then colonizers had perfected their strategies for usurping native land. Not surprisingly, seeding was a prominent tactic. Crosby quotes a leader of a French expedition to New Zealand: "I planted stones and pips wherever I went—in the plains, in the glens, on the slopes, and even on the mountains; I also sowed everywhere a few of the different varieties of grain, and most of the officers did the same." Comments Crosby: "It did not occur to any pakeha [white settler] for decades and decades that spilling and strewing alien organisms into an ecosystem can be like lighting a candle in order to lessen the gloom in a powder magazine."

At the time, much of this seed was dirty—that is, full of weed seed. Thus, the practice of seeding provided a ride for the botanical equivalents of rats and smallpox. The long list of pre-Columbian domestications from the central Mexico Valley gives us some indication of the richness of its native flora; it was a center of biological diversity. Yet, today, an inventory of the region's plants (other than crop plants) produces a list of Old World species. And this takeover had largely been accomplished by 1600, only a century into colonization.

Likewise, although California is regarded as probably the richest center of biological diversity in what is now the United States, assembling a picture of its native grasslands is largely an exercise in paleobotany. California remained isolated and uncolonized until the mid-eighteenth century, yet adobe bricks from as early as 1769 show remnants of curly dock, sow thistle, and red stemmed filaree, all European weeds. That was but a modest beginning. A twentieth-century survey in the San Joaquin Valley found that invaders from Europe and Asia made up 63 percent of the vegetation in grasslands, 66 percent in woodlands, and 54 percent even in chaparral, the scrub brush habitat unique to California.

The same takeover occurred in the East. As early as 1638, a visitor to New England identified and catalogued twenty-two species of English weeds already at home in the colonies. One was indeed dandelion. Another was mullein, a species I battle now on my "wild" land in Montana.

Both weeds and introduced grasses were particularly hard on native grassland communities, a battle that continues to this day but one that was already mostly lost by the natives when it began. The invaders had long before coevolved with grazing sheep and cattle, and so could fill the voids that rampant overgrazing created, a relationship that epitomizes the partnerships that emerged across the advancing face of catastrophic agriculture. European colonization was biological warfare, much of it prosecuted before white eyes were present to record the battles. This strategy emerged full-blown and obvious late in the period of New Zealand, and while whites may not have understood fully what was happening, the native Maoris did. It is worth quoting Crosby's summary of this at length:

In the 1850s, with the avalanche of the pakeha and associated species pouring ashore, more models of Maori extinction appeared. Exotic weeds ran like quicksilver among the roads into the bush. Native birds retreated as exotic cats, dogs, and rats advanced. The inadvertently imported Old World housefly proved to be so effective at driving back the native bluebottle fly, hated by the pakeha because it learned to lay eggs in the flesh of sheep, that herdsmen took up the practice of carrying their own flies along with them into the back country in jars. The brown rats swept through the South Island, again exterminating all but a trace of the Maori rats, and in the 1860s were deep into the Southern Alps and growing to enormous sizes. Julius von Haast, a geologist who arrived in New Zealand in 1858, wrote Darwin that there was a proverb among the Maori that "as the white man's fly drives away our own, and the clover kills our fern, so will the Maoris disappear before the white man himself."

HARD TIMES

Evolution fails to honor terms like "success" and "failure" for very long. In a petri dish, the early relative success of a bacterial colony only yields to a correspondingly early collapse as it approaches the edge of the dish. In the case of agriculture, expansion did indeed build on successful technology perfected in the Old World, but it was also driven by the failure of agriculture in the Old World. Expansion produced excess population, just as centuries of farming overworked and depleted soils. Europe needed a bigger petri dish.

It may be true that Europeans understood their system was flawed only in hindsight, but it was near hindsight. At the time of New World settlement, the life span of the average European was frighteningly short, about forty years—and far shorter in some urban centers (seventeen in Manchester, England, in the mid-nineteenth century, for instance). Transplanted to the New World, these same Europeans began living far longer. Spaniards in the earliest colonies around Buenos Aires reported survivors of up to a hundred years of age. In Massachusetts, the average age at death for the first settlers was 71.8 years. Settlers in New Zealand reported an annual death rate of 5.3 per thousand in the nineteenth century, compared to a contemporaneous death rate in the United Kingdom of 16.8 per thousand. In 1898, for every 10,000 male infants born in New Zealand, 9,033 survived. England's survival rate for the same period was 8,414.

As we have seen, the disease and parasite load the Old World carried played a huge role in this disparity, but something more basic was at work. Settlers in the New World ate better. There are strings of wide-eyed dispatches from European visitors to the New World reporting such tidbits as "feasts" at the boardinghouses of average factory workers, where meat was served twice daily. Factory workers in England during the same period were lucky to see meat twice a month, if at all. The Americans had both more food and better food. Even the temperance crusaders Catherine Beecher and Harriet Beecher Stowe focused more opprobrium on food than drink: "For every reeling drunkard that disgraces our country, it contains one hundred gluttons." It was a novel problem. On one of his visits to France, Thomas Jefferson and a few of his countrymen were enduring a drawing-room assault from a group of Frenchmen touting the virtues of their civilization over that of the New World. Jefferson simply pointed out that the Americans were the tallest people in the room. The European society that produced these settlers was steeped in chronic hunger, and had been since the beginning of recorded history.

The best evidence that hunger was a way of life in agricultural societies is the persistence of famine. Indeed, accounts of famine, like history itself, date to the beginning of agriculture. The food writer Brian Murton goes so far as to suggest that famine is, in a way, history itself: "In preliterate societies, particular famines, along with other collective catastrophes, served as a common means of recording and recovering the experience of the past." In a limited sense, this notion of history probably predates agriculture. Hunter-gatherers like the much-studied Eskimos of the extreme north of North America have a long oral tradition that remembers hunger—such as when a particular village failed in a certain year to catch a whale. They dealt with this as humans have throughout the ages, with a high death rate and infanticide. Yet these are hunter-gatherers on the extremes, living in an environment that most of the world's humans could not tolerate for a single season, even a single week. The better test is in the temperate world, and there, famine is a creation of farming.

Historians have been able to assemble a reasonably complete account of famine stretching back six thousand years. At any given time, this plague was not spread equally across humanity but dominated in a shifting series of famine centers. From about six thousand to twenty-five hundred years ago, the hot spots for starvation worldwide were in northeast Africa and western Asia. This is when Egypt recorded regular cycles of famine lasting seven years, during which, according to an ancient account, "every man ate his children." From twenty-five hundred to fifteen hundred years ago, the Roman Empire recorded at least thirty-five major famines, making it simultaneously the center of hunger and civilization in the world. Then the locus shifted to Western Europe, where it remained until the beginning of the nineteenth century. Between A.D. 500 and 1500, there were ninety-five major famines in England, and seventy-five in France, which suffered its last major famine in 1795. Late in this period, the center shifted to Eastern Europe, which recorded 150 famines between 1500 and 1700. Russia was next, with a hundred major famines from 971 to 1974. In the years 1921–22, about nine million Russians starved to death; in 1933–34, between four and seven million.

Meanwhile, Asia, both north and south, was suffering famine throughout this period, but not of sufficient scale and frequency to challenge Europe's status as famine capital. The incidence, however, rose rapidly in the nineteenth century. South Asia has seen at least ninety major famines in the last twenty-five hundred years, and two-thirds of those came after 1700. At the beginning of the twentieth century, Asia emerged as the world's famine capital. However, China, that long-standing agricultural society, has never been without famine. From the beginning of its history, famine was a way of life. During the past two thousand years, it recorded about ninety famines per century, almost one per year somewhere in the country. Its most devastating famines, however, have been in the twentieth century.

Given this pattern, it is difficult to see agriculture as the antidote to hunger. In fact, the pattern suggests the opposite. Famine was the mark of a maturing agricultural society, the very badge of civilization.

The world's worst famine is not buried in archaeology and the scratchings of forgotten languages. The worst we know of occurred within the memory of middle-aged people alive today, yet its details—the sweep of a period one writer has called a time of "hungry ghosts"—are only beginning to trickle out in anecdote and demographers' calculations. The details have been closely closeted by one of the world's most organized and autocratic societies. Nonetheless, we are coming to understand that, conservatively estimated, eighty million people died of starvation during the period the Chinese call the "Great Leap Forward."

Again, China is the great enigma in the expansion of catastrophic agriculture. Unlike the pre-Columbian agricultural societies in the Americas, China's suite of crops and technologies were the equal of Europe's; in fact, in many cases they were ahead of Europe's, or were Europe's. The northern half of Chinese society was based in wheat, and China had given Europe both chickens and swine. As Europe was gearing up for imperialism and exploration, China was doing the same, sending the eunuch navigator Cheng Ho on a voyage to the Indian Ocean. Chinese trade with Europe during the fifteenth century was widespread. Economists calculate that at this point China accounted for about a quarter of the world's economy. Yet, integrated as it was into the European economy, China never became imperialistic, at least not globally. (Modern Tibetans would consider the distinction academic.) The Han Chinese did finally grab an enormous amount of territory, but it was contiguous.

China had its own internal tensions between north and south that kept it preoccupied. Even more demanding was the centuries-old conflict between this hierarchical agricultural society and the nomadic pastoralists (aka Mongol hordes) to China's west. More fundamentally, China's best agricultural trick was, and is, rice, and rice doesn't travel well in temperate zones.

And yet there is also a case to be made that China tempered its drive toward imperialism by using famine to vent population pressure. Hunger drives imperialism to a degree; Middle Eastern, Roman, and later European imperialism all coincided not only with these regions' development as agricultural societies but also with their accession to the title of global famine capital. Simply put, the population explosion that agriculture allows creates the need for expansion, as it has since the first wheat-beef people hit the European plain. A society, however, also can settle the problem with famine.

The massive toll of starvation during the Great Leap Forward has been read by demographers, just as an asteroid's striking the earth can be read in the crater left behind. During the periods of famine in Europe, local death rates often ran to 80 percent of the population, both from starvation and from the diseases that strike hunger-weakened people. Furthermore, those death rates reflect a higher toll on both pregnant and lactating women, because of their greater nutritional needs. The menstrual cycle is suspended in hungry women, so even those who do survive don't reproduce. All of this combines to level population.

For the last several thousand years, as the famine center has shifted around the world, waxed and waned, China has maintained a fairly steady course of starvation. Researchers have compiled documentary evidence of 1,828 famines in China between 2019 B.C. and A.D. 1911. They were concentrated in a famine belt between the Yellow and the Yangtze rivers, which is to say that China's hunger was concentrated in the same area as its agriculture. This history can be read in the numbers, but also in the language.

During China's most recent famine, people again began using an expression that means "swapping children, making food." Hungry peasants traded children to avoid killing and eating their own. The practice was widespread. The specific phrase for this is about 2,200 years old and meant exactly then what it does now. As the Han Dynasty was founded, in 200 B.C., a single famine killed about half of China's population. The emperor Gao Zu issued an edict permitting people to eat or to sell their children as meat, thus lending legal sanction to a long-established practice. A written report from 2,600 years ago notes: "In the city, we are exchanging our children and eating them, and splitting up their bones for fuel."

The Chinese are not alone in practicing cannibalism during famine. The phenomenon was, for instance, widespread in the Ukraine in the famine of the 1930s. Still, a people that acknowledges cannibalism in tradition, law, and language is indeed a culture of hungry ghosts.

Amartya Sen, in his book *Poverty and Famines*, argues that the problem is not insufficient food but a lack of "entitlement," which he defines as the means to command food. "Hunger and famine have to be seen as economic phenomena in the broadest sense . . . and not just as reflections of problems of food production." Translated from academese, this means people are hungry because they can't afford to buy food, not because there isn't food to buy. It would seem hunger and famine are creations of poverty, not agriculture; but, of course, poverty is agriculture's chief product.

It is an article of faith among modern proponents of the agriculture miracle that famines are a thing of our past. True enough, they acknowledge, famine still does happen in pockets on the globe, but they argue that we produce enough food to feed everybody. That is, famine is no longer a problem of technology but of distribution, not of science but of politics.

One thread in this line of thinking blames modern famine on bad government. China's most recent famine fits this analysis. The immediate cause of the starvation in China during the Great Leap Forward was monumental government stupidity. Specifically, Mao decided that the laws of Western science were invalid in China—a bourgeois plot—and unilaterally decreed a whole new set of agronomic principles drawn from thin air. For instance, wheat and rice were seeded at densities that were orders of magnitude beyond what the soil could support. Crop failure was almost total. Worse, sycophants and functionaries desperate to meet quotas and curry favor up the line refused to report the failures. They faked photos of fields so packed with ripe stalks that peasants could stand on top of the waves of grain, as if walking on water. Party workers fashioned plaster models of giant vegetables as evidence of their miraculous success. Determined to demonstrate this success to the world, the Chinese exported what little grain they had.

We can just as easily trace the blame for contemporary famines in Ethiopia, Bangladesh, and the Sudan—certainly in North Korea—to inept rulers. Modern famine is the result of bad government, but so was ancient famine. Bad government is a part of the syndrome, a chicken-and-egg problem. Population explosion generates the need to grow more food, but agriculture is the cause of that population explosion, and agriculture creates government. The hierarchical, specialized societies that agriculture builds are wholly dependent on the smooth operation of their infrastructure, on stability, on transportation. Dams must be built, canals must flow, roads must be maintained, and government must be established to order those tasks. Government leaders emerge from the social hierarchy that agriculture's wealth makes possible. Failures occur as frequently as humans fail. To hold agriculture blameless and government responsible for famine is like holding a lion blameless for a child's death on grounds that it was the lion's teeth that did the biting. Poverty, government, and famine are coevolved species, every bit as integral to catastrophic agriculture as wheat, bluegrass, smallpox, and brown rats.

The food historian Sophie Coe has argued that famine did not begin to end in Europe until the introduction of New World crops. Much has been made of the effects of Spanish ships returning loaded to the gunwales with gold and silver, and much should be. The loot reorganized not just Europe's but the world's economy. China, for instance, found itself heavily drawn into the European trade network in the fifteenth century simply because most of its currency was made of New World silver. The flash of this booty, however, burned out fast. Silver and gold do not go forth and multiply; but seeds do. Long after Spain's gold—and Spain itself—faded from prominence, maize and potatoes remain. Maize is today the world's most important crop, potatoes fourth. This was then and is now of special significance to the world's poor. Most of the maize grown feeds livestock and poor people, and this low-caste bias is even more pronounced in the case of the humble potato. While Europe's upper crust learned to sip chocolate and munch tomatoes, the potato reorganized the life of Europe's peasant—and later its working—class. It's worth tracing that trail for what it tells us about food's ability to sort social classes.

The potato was not an overnight success. Still, the conquistadors did recognize its potential when they found it in the Andes no later than 1537, along with a string of indigenous domesticated tubers that are still central to Andean agriculture. They had ferried potatoes back to Europe by 1570, but it would be two centuries before the plant developed into one of Europe's most consequential foods. That dormant period had everything to do with where the first potatoes were found.

Andean agriculture is the exception to the rule. That is, all of the other agricultures in the world—wheat in the Middle East, rice in Asia, maize in Mexico—sprang up in fertile river valleys close to temperate zones. Andean agriculture developed atop some of the world's tallest mountains, close to the equator. Thus, the climate in which potatoes evolved was relatively cool, because of the altitude, but with a long, light growing season, because of the latitude. They are photoperiod-sensitive, which is the botanist's term for being sensitive to the length of day. That is, they take their key signal to shift gears, from building roots and stems to building tubers, from the seasonal change in the length of days.

As a result, the first potatoes brought back from the New World wouldn't grow well in much of Europe. They required a very long growing season, and would only set tubers in late fall, well after temperate-zone frosts killed the vines. There was an exception, however. Northern Atlantic currents give the British Isles a mild fall climate. The Andean tubers could thrive there. Later, an American breeder developed some Chilean potatoes into varieties more suited to a temperate zone. These are the types that dominate production today and made the spud at home throughout Europe in the nineteenth century. Ireland, however, got an early start, a classic case of agricultural success prefiguring doom.

To a certain degree, England, too, could have benefited from potatoes, as Ireland did. The climate agreed with them, but the social atmosphere did not. Throughout Europe, there persisted a vicious prejudice against the potato as a food fit only for livestock and the Irish. The French, for instance, regarded the potato as poisonous. In the nineteenth century, when Irish immigrants began bringing potatoes into English working-class towns, social commentary railed against the corrupting influence of this food that promoted "idleness, improvidence, and moral deviations." In his book *The Potato: How the Humble Spud Rescued the Western World*, Larry Zuckerman writes:

> The indictment was plain. The potato, a coarse food that subverted desires for comfort or cleanliness, stood accused of cheapening lives. Not only did it promote the ruinous cycle of poverty and population, it was partly responsible for the moral illness that helped bring about tuberculosis, typhus, and cholera.

That is, it was not poverty that made those Irish immigrants eat the potato but their eating the potato that made them poor, sick, and dirty. Not that life was all that great for England's native poor. The disparagement of the potato coincided with that previously reported period when the life expectancy in Manchester was seventeen years. The poor in Britain—which is to say most people in Britain—ate mostly bread. Period. Occasionally there was milk or cheese; meat appeared only on special occasions. Further, bakers stamped their loaves with a "W" or an "H," indicating "wheaten" or "household." The pure wheat bread went to the well-to-do. Bread of mixed wheat and rye flour, the household bread, went to the poor. This thin diet based wholly in grain left the British terribly vulnerable to crop failures and the ensuing famines, a situation that really didn't change until Britons reluctantly adopted potatoes in the nineteenth century. That, and they began exporting starvation to colonies.

The Irish adopted potatoes earlier because they didn't have much choice. Wheat bread is a simple enough diet, but still requires some resources: the grain must be ground and ovens must be fueled to bake it into loaves. Potatoes are the food for the poorest of the poor because they don't require even these steps. One digs them from the ground, a dense package of starch that needs no milling but can simply be tossed into the fireplace to roast and then be eaten like an apple, no fork or plate required. This ease of cooking was critical, because England had clear-cut Ireland's forests, and, unlike England, Ireland had very little coal. Fuel was peat, and eating potatoes instead of bread helped to conserve that fuel. Spoiled spuds could be tossed to hogs, allowing the poor to raise a bit of meat (for market, of course, not for table).

Said the Irish, the "sauce of a poor man is a little potato [eaten] with a big one."

Before potatoes arrived, the Irish relied heavily not on wheat but on porridge of oats, a crop bred to grow on the ragged edges of agriculture where wheat won't. Ireland could grow wheat on its eastern edge, but as with beef, only a few well-to-do Irish would taste it, and for the most part it was exported to Britain. The island's pitiful, rocky soil, its cool climate, and frequent rains all combined to make growing wheat a difficult assignment, so oats would have to do (and even those failed frequently). The rain-tolerant potato changed this situation rapidly. By the beginning of the nineteenth century, the average Irish person was eating five and a half to six pounds of potatoes per day.

Throughout the eighteenth century, there was a clear trend toward potatoes, accelerated by a couple of major famines. The latter and worst of those occurred in 1740–41 and killed as many as 400,000 people, 10 percent of Ireland's population. Those famines were brought on by failure of the oat crop, prompting an increasing reliance on potatoes. At the beginning of the period, the poor—the majority of the Irish—were living on milk, potatoes, and oat porridge. Toward the end of the century, the porridge disappeared. One observer reported in 1780 that an Irish person usually ate only milk and potatoes year-round "without tasting bread or meat. Except perhaps at Christmas once or twice." Meagerly equipped kitchens meant the potatoes were simply boiled and eaten. People commonly kept one fingernail long to strip the peel, the only cutlery required.

In the meantime, potatoes had been the basis of Andean agriculture for thousands of years. Yet there it built a culture of such wealth that the Spaniards came to dominate the rest of the world simply by looting the Andes. So far as we know, though, the Incas were never in the same position as the Irish. Perhaps they learned the hard way the perils of depending on a single crop. For whatever reason, the diet in the simplest Andean village was (and remains) far more varied than that of the peasants of the advanced Western world. The Incas grew maize, a string of indigenous tubers (especially mashua, ulluco, and oca), and quinoa (a species of lupine) for grain. They had vast numbers of llamas and alpacas, and the rich ate both. The peasants, however, were not without protein; they ate guinea pigs, a delicacy that still figures prominently in Andean peasant diets. It was not a monocrop economy, but its mainstay, the potato, created a monocrop economy in Ireland.

But for all of its advantages, the potato has one enormous drawback. Because of its complex array of chromosomes, it doesn't reliably pass on traits from generation to generation, or even regularly produce seed. Cultivation depends on vegetative propagation. As any backyard gardener knows, one does not grow potatoes by planting seed but by planting a bit of last year's potatoes. Each new crop is a clone of the last. If you had the sensibilities of a pathogen, you would see enormous opportunities in this. Bits of tissue carry on, generation after generation. Normally viruses and fungi aren't so lucky. They can't survive in most seeds, and have to find alternate hosts to weather out the off season. With potatoes, viruses and fungi go along for the ride year after year.

At the time of the Irish potato famine, though, viruses were not on European minds, people were, and the focus was on Ireland. The potato's "success" could be quickly measured and was apparently evident, even to the Irish of the time. Typical households had six to ten children. A French visitor once asked an Irish family how such poor people were able to raise so many healthy children. The answer was, "It's the praties, sir."

The economics of rural life stabilized, allowing earlier marriage. Above all, the hand-to-mouth existence of the countryside prizes plentiful, cheap stoop labor—i.e., children, an asset made even more abundant by Ireland's Catholicism. As a result, between 1780 and

1841 the population of Ireland doubled from four to eight million. It was simply a population explosion, and English observers, many of them infected by the vicious anti-Irish sentiments of the time, noted it warily. In fact, this explosion caused the father of gloom and doom, Thomas Malthus, to focus his studies on the potato and to speculate that Ireland would soon have a population of twenty million. From our vantage, Malthus's projections have often been ridiculed as unnecessarily grim and, in any event, wrong. They were wrong, though, only because Malthus didn't account for factors like fungi. The devastation wrought by plant disease balanced Malthus's equations, a "correction" that was hardly cause for optimism.

The unprecedented boom in population and the country's dependence on a single, disease-prone crop made Ireland doubly vulnerable in 1845. The fungus in this case (*Phytophthora infestans*) leads to a disease called late blight. It is still with us, and is still the leading cause of potato-crop losses around the world.

Phytophthora infestans first made its way from North America into the Low Countries of Europe in the 1840s. Because those places were not nearly so dependent on potatoes as Ireland, the spread of the disease and the resulting losses were moderate on the Continent. When it appeared in Ireland, in 1845, it claimed about 40 percent of the crop that year. Late blight causes potato vines to curl, blacken, and die. The diseased tubers are inedible, so they were simply left to rot in the ground. This practice allowed the fungus to weather the winter of 1845–46, then roar to life the following growing season. Ironically, it was aided by earlier fungal and viral diseases. European potatoes had suffered before from a different fungal dry rot and a viral curl, and as a result growers had planted varieties developed in North America that were resistant to these diseases. This narrowed the number of varieties grown in Europe, and it happened that these new varieties were wholly susceptible to late blight. In 1846, weather conditions were ideal for the spread of late blight, and spread it did, knocking out 90 percent of the single crop on which eight million poor people, most of them children, depended. The blight abated slightly in 1847, then came back full force in 1848. As often happens with famine, disease—in this case, cholera—spread through a weakened population. The usual companion of hard times, government stupidity (or viciousness), followed closely. The English refused any amendment of the corn laws, which meant that a starving Ireland continued to export grain to England.

The commonly quoted toll of the Irish potato famine is a million people. Another 1.3 million emigrated during that same period. Over the next sixty years, another 5 million fled the famine-ravaged economy. By 1911, Ireland's population was 4.4 million, about what it had been in 1780, before the potato-induced explosion.

Dehumanization probably took as great a toll, if a less measurable one. An observer reported that the "bonds of natural affection were loosened," that parents neglected children, and children parents, and men abandoned wives and children. As Zuckerman notes, one Irishman reported a visit to a village on the western coast, where the famine was worst:

> In the first [hovel] six famished and ghastly skeletons, to all appearances dead, were huddled in a corner on some filthy straw, their sole covering what seemed a ragged horse cloth, and their wretched legs hanging about, naked above the knees. I approached in horror, and found by a low moaning they were alive, they were in fever—four children, a woman, and what had once been a man. It is impossible to go through the details. Suffice it to say, that in a few minutes, I was surrounded by at least 200 of such phantoms, such frightful specters as no words can describe.

England overcame its hatred of potatoes, not necessarily by choice, and certainly not quickly. By the beginning of the nineteenth century, some Britons were beginning to replace their beloved wheat with this cheaper, more convenient starch, but no one really liked the idea, particularly not the upper class. One writer, calling potatoes "the root of slovenliness, filth, misery, and slavery," concluded that he would rather be hanged than eat "the lazy root." The potato stood as chief target for the proper Englishman's hatred of the Irish in particular and the poor in general. England, however, depended on the existence of a vast stock of poor people, and so, by extension, needed the potato.

Enclosure, a practice that had impoverished Ireland by excluding subsistence farmers from farmlands taken for the aristocracy, also occurred in England in the early nineteenth century. Landowners consolidated small peasant holdings, reducing the peasants to wage workers. These newly landless workers had no fields to grow grain, but a hill of potatoes occupies only a few square feet. Thus the potato, the backyard food source, came to fit into their diet. The potato's means of preparation, however, were even more relevant than the means of growing it.

Wage earners began to congregate in the infamous slums of industrial England, where tenements lacked any sort of cooking facilities. In fact, it was standard practice for a working-class family to rent oven space on Sunday, their day off, to bake a bit of bread. Only a generation before, rural workers had taken something like 40 percent of their nutrition from bread alone, but in the slums, as industrialization gained steam, bread became a luxury. It was replaced by the potato, which could be roasted and eaten on the street. A number of institutions grew from the necessities of this new, portable life. For instance, it was the practice at many factories to hand the mass of workers a lump sum on payday; it was up to them to make change and divide it amongst themselves, and public houses arose for this purpose. In the process of making change, the public house would hold on to a bit in exchange for a pint or two. These same workers— and the family members who came to drag them from the pubs— could immediately trade some of the cash for street food, fried potatoes and sometimes fried fish. The combination is a marriage of convenience that survived as fish-and-chips and puts the lie to the notion that fast food was born in mid-twentieth-century America in a phalanx of deep fryers nestled under a set of golden arches.

The wheat-beef people who first confronted the Hungarian plain six thousand years ago created an agriculture that would sweep the world. By this measure, their culture was successful. Yet by another measure, that culture had no choice but to sweep the world, being, as it was, the engine of population growth that would so alarm Malthus in the nineteenth century and Paul Ehrlich et al. in our time. The New World provided a safety valve, but so did famine and disease. A million dead in Ireland—and more in earlier famines in England, France, Russia, and Eastern Europe—adjusted the demographics to a more manageable level, as did the hunger-driven emigration of fifty million people from Europe, largely to the neo-Europes, between 1820 and 1930. The introduction of New World foods to Europe also offset some need, allowing population to swell further. These new foods came just as Europe was industrializing, just as it needed a vast, cheap source of labor and new, cheap ways to feed it. The potato filled that role, but there appeared at the same time another new tool for concentrating populations of the poor.

As much as spices and slavery, sugar drove European exploration and imperialism. Sugarcane, a perennial grass, was first domesticated in Southeast Asia. During the Arab agricultural revolution in the Middle East, which coincided with the rise of Islam, it made it to Mediterranean Europe, arriving in Spain with the Moors. The Iberian peninsula was the ragged edge of the tropical grass's natural range, but the Spaniards were able to grow it, a skill that would give them a leg up in the contest of imperialism. Even before Columbus, the Spanish developed colonies at Madeira, and then in the Canary Islands, and quickly discovered that both of those places grew sugar well. The locals, however, did not. The indigenous people of the islands were hostile and proved to be unreliable laborers, largely because they died from European diseases. The Spaniards solved this problem by importing slaves from agricultural areas of Africa, a practice that would dovetail nicely with their new expertise at growing sugar. This initial foray, and the resulting trade in sugar with the rest of Europe, financed Spanish exploration. The model built in the Canaries set the stage for what food historians call a "sugar revolution" in the eighteenth century. Brazil and the West Indies provided ideal sugar climates, and slave traders had perfected their methods to allow the importation of ten million African slaves to the sugar colonies. This in turn created a glut of sugar, so what had been a curiosity, a luxury for sweetening the ladies' chocolate in Europe, suddenly became commonplace and cheap. As this happened, the British came to dominate both the trade in sugar and its consumption, by virtue of the territory they held as colonies and their dominance of trade in Senegal.

The spread of slavery, the sugar trade, and the consumption of sugar powered the Industrial Revolution, for reasons particular to the nature of sugar. First, sugar agriculture is, by necessity, tropical. It is not a simple extension of Europe's wheat-beef base, which wouldn't work in the tropics (nor would the wheat-beef methods). Temperate agriculture could spread with the Jeffersonian yeoman farmer model because Europe was filled with trained farmers, but the farmers of Europe grew food that required, at most, grinding and cooking. Sugar required a labor-intensive industrial refining that made it very much the first processed food—processed by slaves. And sugar was a remarkably efficient food, producing the most calories per acre of any crop. An acre of sugar will produce the same number of calories as four acres of potatoes, twelve of wheat, or 135 devoted to raising beef.

There is a fundamental tension inherent in civilized economies, one that intensifies as they develop. Farming, pyramid building, and industrialism above all require a huge pool of cheap labor. But that pool must be fed. We have seen that famine, disease, and simple malnourishment can come to the rescue of an overtaxed economy by correcting periodic population imbalances. In this light, famine, poverty, and disease are useful institutions, which is perhaps why Christ was so certain they would always be with us. The more trumpeted tool for this task, though, is efficiency, a favorite word of economists. In this mind-set, food is no longer a pleasure, an aesthetic experience, a bearer of culture and tradition. It is not cuisine but calories. The efficiency of sugar fit nicely with the ascendant dehumanization that was British industrialism.

Sugar gave the homeland cheap food, supported by slave labor in the Caribbean and South America. Its production rested on industrialized plantations that were markets for England's factories. The plantations in turn created wealth that became the capital that financed the industrialization of Britain. It was a system that had nothing to do with the well-being of most of the humans involved and everything to do with raising wealth. Writes the anthropologist Sidney Mintz, "Slave and proletarian together powered the imperial economic system that kept one supplied with manacles and the other with sugar and rum."

The British custom of taking tea as an afternoon break has more to do with sugar than with tea. During the nineteenth century, when the custom arose, it was something like the coffee break in modern workplaces, but not so leisurely: a chance to gulp a quick cup of tea, which was invariably laced with sugar. In this way were the human machines of the factory "nourished"—fueled—without even needing to leave their machines.

Annual per capita sugar consumption rose 2,500 percent in England during the 150 years preceding 1800, but the pace quickened as the nineteenth century brought a proliferation of not only sugared teas but also jams (which were about 60 percent sugar), puddings, and treacle to British tables, especially those of the working class, resulting in a 500 percent increase in consumption in the years 1860 to 1890. By the beginning of the twentieth century, the average Briton was getting about one-sixth of his total nutrition from sugar.

Mintz argues that these figures don't tell the whole story, in that dependence on sugar was not only greater among the poor, but among women and children in working-class households. Any meat and bread went to the man of the house as a simple practical measure: the mister needed his strength to work. Women and children typically ate jam, suet pudding with molasses, and sugared tea for at least two meals a day. Says Mintz, "If that figure [one-sixth of calories] could be revised to account for class, age, and intrafamily differentials, the percentage for working-class women and children would be astounding." This is efficiency too, if one considers the purpose of humanity to provide cheap labor. Like famine, malnutrition promotes infant mortality and suppresses the birthrate, biasing the population toward working adults.

The rationale remains with us in the form of an order of fries and a Coke. The spirit of fast food in twentieth-century America is an echo of the nineteenth in Britain. Fast food got one of its bigger boosts when Bill Clinton, then president, confessed an affection for Big Macs. More than just a politician's slumming, this presidential touch lends the legitimacy the system needs. Similarly, fish-and-chips had maintained a nasty reputation in Britain into the twentieth century, but much of that was overcome when Winston Churchill confessed a secret liking for it, thus providing state sanction of a necessary institution.

MODERN TIMES

There is a story in my family about my paternal grandfather, a respected and successful, albeit bullheaded, farmer in northern Michigan. I heard it from my father, who was not a farmer—nor were any of his three brothers who survived to adulthood. They left my grandfather's farm as soon as they were old enough; they'd seen quite enough of backbreaking labor, the chief product of my grandfather's three hundred or so rocky acres in Alpena County. Instead, they went to southern Michigan's cities to swing sledges and push wheelbarrows full of concrete, a considerably softer existence. Still, I was raised on the backside of the glacial escher that formed the highest hill in Alpena County, Manning Hill, as was my father and his father before. A silo that bears my great-grandfather's initials and a date, 1919, still stands.

My grandfather was notorious in his community, known for his acumen as a farmer and horse trader, but mostly for his obstreperousness. I was afraid of him, like everyone else, afraid especially that he would take off his glasses and expose what lay behind the one frosted lens, the empty socket of the eye he had lost in a farming accident. I was less afraid, though, after the day when my parents took me to show him an exemplary report card and he pulled a dollar bill from his seldom-opened wallet and gave it to me, with the advice that I get an education so I wouldn't have to farm. No one could argue with his assessment that his was a tough life.

The story my father told me about him occurred during the Great Depression, in a period of poor prices. Then, my grandfather raised mostly potatoes. That fall, he loaded a truck full of potatoes and took them to the local selling shed, where buyers offered him a price he thought pathetic. So he refused to sell, backed the truck across the road, dumped the potatoes in the ditch, and then drove the truck over them to crush them, as the buyers looked on. To this day, farmers are offered pathetic prices for crops, but no one in his right mind would do what my grandfather did.

As far as I know, he was in his right mind, and besides his potatoes, he also had at home cattle, hogs, chickens, eggs my grandmother used or sold, milk and cream from cows, apples, seed potatoes saved, and manure piling up to fertilize next year's crop. A wood lot gave him lumber and fuel to heat the house. Neighbors supplied him with labor when he needed it, and he repaid them in kind. He had alternatives, and could get through a year without selling his potatoes. His was the last generation of farmers to have that independence, before it got traded away for efficiency.

My grandfather probably got about the same yield per acre on the wheat and oats he grew as did his forebears—respected, bullheaded British wheat-beef farmers—three generations before. They, in turn, probably produced roughly the same yield per acre of wheat as had their Roman predecessors two thousand years before. There is some evidence that the Romans, in fact, harvested poorer yields than had the Sumerians four thousand years ago, the decline a result of Italian soils depleted from generations of farming. But regardless of minor fluctuations, virtually all of the increases in total food production, from the advent of farming to my grandfather's time, were achieved by expansion of the arable land base, which is to say, by bringing more fields and more continents under the plow. Eventually, though, expansion ran up against the limits of the planet's supply of plowable land. From that point, almost all the increases in total food production have had to be achieved by increasing yield—by harvesting more bushels per acre. This shift set the terms of modern agriculture, and is almost as significant as the development of agriculture itself.

This shift in the fundamental benchmark of farming occurred about 1960, just after my grandfather quit farming and a year before he died. Strange as it may sound, that year stands as a halfway point in the development of human culture. Lloyd T. Evans, in his excellent history of farming (*Feeding the Ten Billion: Plants and Population Growth*), divides time by the billions of people. That is, a certain set of agricultural and cultural techniques grew up as the living human population grew to one billion people, which occurred roughly in 1825, then nearly ten thousand years into the agricultural experiment. The second billion would take but a century to add, and we hit it in 1927; my grandfather was already on his farm by then. In 1960 we hit three billion, which is about half of today's population and, hence, a halfway point. Of course, "halfway" is relative to where we are headed, so, to a degree, the selection of that year is arbitrary. The shift from increasing acreage to increasing yield at almost precisely that point, however, is not arbitrary. It was allied with a number of developments that we would call collectively the Green Revolution, at least when it became international. The foundation of this revolution, however, took shape in the world's foremost agricultural nation, the United States, during the era of my grandfather's tenure on his farm.

The shift did not occur overnight, but only after a half century of ominous warnings and catastrophic failings plainly spelled out the limits to agricultural conquest of arable lands. When I was a kid growing up on the same land my grandfather had farmed, it was covered with a mix of bromegrass and clover—a simplified copy of a prairie's design—its most important attribute being that it was perennial. My grandfather had retired this land from farming. By law, it could not be plowed, because it had been enrolled in the federal land-bank program, a development that set my conservative Republican grandfather to ranting about the advance of socialism and such, but he enrolled in it nonetheless. By the turn of the century, the same sort of ground cover presided over 36.4 million acres of land across the United States, land enrolled in the modern counterpart to the land bank, a federal plan called the Conservation Reserve Program that pays farmers not to grow crops. These programs came as a result of hard lessons.

Nature began teaching these lessons in the first part of the century, particularly around World War I. Then, Europe's turmoil made wheat crops so valuable as to encourage the sort of speculation we would associate more with tech stocks and Silicon Valley than with plows and dirt. City people—bankers, shoe salesmen, and such—became what were known then as "suitcase farmers." They would rent or settle a section of unplowed land, hire somebody to bust it up and plant a crop of wheat, then retreat to the city and wait for harvest's riches. The phenomenon was about as sustainable as the tech-stock boom.

The problem was that many of these lands had gone unplowed for a reason. They were marginal, which is to say, hilly, arid, and prone to erosion. By virtue of the stored natural capital that thousands of years of grass cover represented, they could raise a good crop for perhaps two or three years, but the stored organic matter and the moisture it held depleted quickly. And when it did, the speculators left, and the land went to dust.

These incursions onto marginal land were not isolated events, but occurred in waves, in booms and busts, throughout the twentieth century. We would like to think dust bowls are bits of history relegated to jerky newsreels, Tom Joad, jalopies, and Woody Guthrie, but in the 1970s and 1980s the lessons of the Dust Bowl proper were forgotten by federal farm policy, and there was a series of great "plow-ups," leading to periodic repeats of the original. There is more to be learned from this than the fact that institutional memory is short and that profit aids amnesia. Throughout the twentieth century, population pressure was pushing farming into marginal lands, places so arid and with soil so thin as to be able to sustain crops only sporadically. This phenomenon was by no means restricted to the United States, but was occurring in Canada, Australia, and, especially, the former Soviet Union. The Russians, just like the Americans, were at the time engaged in an intense program of industrializing farming that brought an additional thirty-six million acres of the Central Asian steppes under cultivation.

Evans has calculated that, in all, the arable land base grew from 1 to 1.4 billion acres between 1928 and 1960. During the same period, though, a great deal of land once dedicated to feeding draft animals was switched over, with the advent of the tractor, to feeding humans. With these lands factored in, the period saw a 53 percent increase in arable lands, just as it saw a 50 percent increase in global population. The most dramatic sign that we had gone too far was registered by nature during the Dust Bowl. A series of drought years beginning in the mid-thirties—and drought years are inevitable in the plains—parched the landscape; then winds raised a cloud of dust that literally blackened the skies half a continent away. The cloud, in fact, shadowed Washington, D.C., just as Congress was considering measures to deal with the erosion, probably the only example we have of nature lobbying so directly on its own behalf. We had to pull back; but, more important, we had to devise a new strategy to feed added population.

The year 1960 would be pivotal, but fundamental change had been percolating through the system before then. In fact, corn yields had already begun to rise by the 1940s in the United States, a sort of mini-revolution that presaged all that was to come.

The root of that rapid increase in corn yields was hybridization. Unlike the human experience with wheat and rice, the process of domesticating maize was long and arduous. Archaeological remains of early domesticates show ears of corn the size of a grasshopper, hardly the fat clump of food we have come to expect. Careful selection over hundreds of years finally produced a big enough seed head to support temples, potentates, and libraries. From then until the advent of hybridization, however, the productivity of the crop seemed hedged by inherent limits.

Hybridization is defined simply as a wide cross, or breeding unlike mates from different strains. Corn is open-pollinated, which means varieties naturally cross, so a certain amount of hybridization always occurred, but breeders concentrated on pure lines, crossing closely related strains with similar traits. No less a figure than Charles Darwin first suspected the benefits of deliberate hybridization in 1877. He and a colleague, W. J. Beal, reported a characteristic that would eventually be known as "hybrid vigor," meaning the progeny of these wide crosses outproduced both parents. Curiously, though, they did not pass on that vigor to succeeding generations, so the process appeared to be a dead end until well into the twentieth century. The Wallace family of Iowa was instrumental in reviving and furthering the development of hybridization. F. D. Richey, an early experimenter with hybrids, joined the staff of U.S. Secretary of Agriculture Henry C. Wallace, and the USDA began promoting hybrids. Wallace's son, Henry A. Wallace, developed and began selling hybrid seeds in 1922, founded Pioneer Hy-Bred Corn Company of Iowa (still a major player in the field) in 1926, and then became Secretary of Agriculture, and eventually vice president, under Franklin Roosevelt.

Not since the days of the Founding Fathers did this country have political leaders so wound up in revolution. Agricultural research in the United States today is peppered with gray eminences, all at about retirement age or just beyond. Many of them have made significant contributions to agriculture worldwide. They share a common story, and hybrid seeds are to them a personal matter. They were raised on midwestern farms and remember a dirt-poor existence; then the advent of hybrid seeds, a surplus, and finally enough money that they could think about going to college. They would need to; all of a sudden, farming became a more complicated business. Hybrid seeds shook the foundation of American agriculture almost overnight.

The revolution was swift. Hybrids accounted for about 1 percent of all corn planted in the United States in 1933, and 50 percent only ten years later. Those were the Depression years, years of desperation and drought. Recalls Don Duvick, one of today's gray eminences who worked as director of research for Pioneer, "In 1936 the rains failed again, suddenly in July, accompanied by a terrible heat wave. We got no corn again, except on five acres planted to some expensive seed, something new called 'hybrid' corn, sold by the Jung Seed Company in Wisconsin. Neighbors came from miles around to see our hybrid corn." It raised enough money for Duvick to go to college.

On the surface, the change was marked by a rapid consolidation of farms, first in the South, then in the Midwest. Sharecroppers were quickly displaced; farms became larger and less diverse, principally because of economies of scale, but also because of uneven adoption. That is, some farmers began growing hybrids earlier than others, prospered, and quickly bought up the land of less progressive neighbors. At the same time, farming became more capital intensive. Manure from livestock became less important as fertilizer; farmers began buying chemical fertilizers and, to a small extent, the first pesticides, particularly two—2,4-D and DDT—that would become notorious. Pesticides entered the equation not only because they were just then developed and so were newly available but, more important, because the increasing corn crop, grown from a very few varieties, made a bigger and easier target for pests.

Partly, this shift to a capital-intensive agriculture stemmed from the unique nature of hybrids. Because hybrid vigor does not pass to progeny, farmers could no longer save some of one year's crop to seed next year's. Producing seed became a separate business, and each year, farmers would have to buy what would eventually become known as an "off-farm input." This subtle change signaled the integration of farming into a host of industrial processes. Fuel for plowing was no longer farm-grown hay fed to farm-bred horses, but store-bought fuel fed to factory-built tractors. Fertilizer came from chemical plants, not the floors of stables and corrals, now emptied of horses. Before, farming had been uniquely autonomous of industry, because machines couldn't make food, only nature could. All of a sudden, machines were integral to the process.

Agriculture has always required a subsidy, which is the human-caused disturbance of natural systems, in order to flourish. Its bias toward catastrophe must be maintained with human energy and motion toward the frontier, forever simplifying and subverting natural systems. In the decades leading up to 1960, however, this intervention took on the trappings of industrialization. In doing so, it spread the footprint of farming off the farm for the first time. That is, having run out of arable land, farming in effect began to claim oil fields, steel mines, phosphate mines, and the network of gravel, steel, and asphalt needed to connect them. Once farming ran out of arable land to devour, it started in on the rest. Its consumption did not cease in

1960; technology simply enabled it to begin chewing up landscapes it had once been unable to digest.

Or to turn the metaphor in another direction: a new species (industrialism) emerged into dominance in the nineteenth century and, over the course of a century or so, coevolved, even hybridized, with the old dominant species (agriculture). The resulting hybrid may be sterile, incapable of nurturing future generations, but what it lacks in sustainability it makes up in vigor and tenacity.

By 1960 it was clear that agriculture's historical strategy of expansion was exhausted, but this was by no means the end of its imperialism. Rather, the end of one era ushered in the next. What came next is commonly called the Green Revolution, but it might just as well be called the Age of Dwarfs. True enough, increases in corn yield had been achieved with hybrids, but that strategy offered less promise for the world's other two staples, wheat and rice. These crops are self-pollinators, which makes the process of effecting those productive wide crosses much more difficult. Pollen from a plant's own seed head gets the job done before the alien stuff can invade. Still, if there could be no more acres for wheat and rice, then there would have to be more wheat and rice seed produced per acre. This imperative pointed breeders to the plant itself. Thus began a new kind of biological imperialism, whereby one part of the plant would be usurped by another. Breeders began to redesign what they call a plant's "architecture" to accommodate industrial efficiency.

A plant gets only its place in the sun, so its growth is limited by the amount of sunlight striking it. Plant breeders, though, figured out that people could appropriate more of that energy for the seeds, the part we eat, by taking it away from the stems, and so they bred dwarfs. There was a secondary benefit to this approach: breeders knew that the taller the stem of a plant the greater the leverage exerted by the heavy seed head on the plant's support structure, resulting in a problem called "lodging," whereby the seed head topples the plant and grain is lost in the dirt. Farmers were limited in the size of the seed head they could breed for, and in the amount of fertilizer they could apply. Too much would mean too fat a seed head, which would collapse its stem. The shorter plants produced by dwarfing allowed heavier heads. Just as the chemical fertilizer industry appeared on the scene, ready to send farmers massive amounts of nitrogen, phosphorus, and potassium, agriculture provided plants that could bear the load.

The breeding of dwarf varieties was not new. Visiting Japan in 1873, the U.S. Commissioner of Agriculture, Horace Capron, noted, "The Japanese farmers have brought the art of dwarfing to perfection." As Evans points out in his book, Capron observed that the wheat ears were heavy, but borne on such short stems "that no matter how much manure is used . . . on the richest soils and with the heaviest of yields, the wheat stalks never fall down and lodge."

Meanwhile, English farmers had been working with a dwarf variety in the mid-nineteenth century. By 1917, Americans were crossing Japanese dwarfs with American varieties, a process of selection that continued into the 1950s. Norman Borlaug, trained as a forest pathologist, was then working in Mexico, not on dwarfing, but on a wheat disease. In 1954, someone sent him some dwarf seed, which he crossed with enormous success with a Mexican variety grown in the Yaqui Valley at the edge of the Sonoran desert.

It was indeed a revolution. By 1970, dwarf varieties occupied a quarter of the total wheat area in the developing world (excluding China, which was, as we have seen, then off in its own world). That figure would reach 40 percent by 1975, and greater than 70 percent by the turn of the century. Meantime, breeders quickly repeated the dwarf trick in rice. Dwarf varieties occupied about 40 percent of the land sown to rice in South and Southeast Asia by 1980, and 74 percent by 1990.

In just a single eleven-year period, 1975 to 1986, rice yields jumped 32 percent worldwide, wheat yields by 51 percent. Coupled with gains made earlier from corn hybridization, these quantum leaps created a technical and social revolution in the United States, especially in the heartland. U.S. corn yields stood at about twenty bushels per acre in 1900, as they had more or less through all of history; but by century's close, they had jumped to more than 130 bushels.

Agronomists have another way of measuring this, a number called the "harvest index," which is simply the weight of edible grain compared to the weight of the total plant, calculated as a percentage. Using British wheat as a standard, we can gauge the effect of the revolution. The index in Britain was about 35 percent in 1920, probably a number that held through most of history. Modern varieties of wheat and rice have a harvest index of 50 to 55 percent, a measure not only of the past but of the future of this dwarfing strategy. Agronomists think that the maximum harvest index possible is somewhere around 60 percent: a plant needs to devote at least 40 percent of its production to growing its infrastructure of leaves, stem, and roots. In other words, we have nearly exhausted this strategy in about forty years, a period in which the world's population has doubled. There is an odd proportionality to that. It took ten thousand years to exhaust the old expansionist strategy of claiming more arable land, a period that added about three billion people to the total living population, about the same number added as we exhausted the second strategy.

Harvest index is a measure of efficiency; even more than dwarfing or chemical fertilizer, the pivotal change in agriculture in our time has been the use of this word. Dwayne Andreas, the legendary patriarch of agribusiness giant Archer Daniels Midland, likes to tell a joke in interviews. He customizes it to his own experience, as joke tellers do, but it's actually an old bit of Irish humor. In his version, Andreas is the protagonist, and claims the story arose in his early career, when he made his way across Iowa trying to sell feed supplements to hog farmers. Farmers, at least in popular opinion, are practical and thrifty folks, so they wondered why they needed these new products. Andreas told them the supplement would allow them to take their hogs to market at least two weeks sooner. "But what's time to a hog?" the farmers replied. Andreas laughs when he tells this story, probably because he knows that farmers who think this way are all gone.

The success of the high-yielding varieties of the three main grains, along with attendant technologies like chemical fertilizers, pesticides, and sophisticated machinery, exploded the production of those crops at the expense of all others. That enormous increase in production torpedoed prices, further squeezing farmers for more efficiency. Many of them went broke in a series of farm crises that continue to the present. Successful farmers then bought up the land of the unsuccessful in a drive for economies of scale, allowing them to pump out still more grain. Now, no farmer would dump his crop as my grandfather did rather than accept a pathetic price, because the government makes up the difference. This is one of the legacies of the drive for efficiency.

As with dwarfing, there is nothing new about crop subsidies. Ancient Rome paid them to its farmers. Still, it would be difficult to imagine any other time in the long history of agriculture when farmers would admit, as they often do now, that their primary skill is "farming the government." The United States has been in the subsidy business since at least the nineteenth century, although the older forms are more indirect. President Lincoln, for instance, created the Land Grant College system, the Homestead Act, and the railroad land grants, all of which amounted to enormous subsidies for settling and farming the Midwest and West. However, the business of directly supporting prices with cash payments to farmers didn't emerge in force until the Great Depression. These payments were intended to keep food cheap, to keep farmers on the land, to diversify the crop base, and, above all, to be temporary. They have succeeded only in the first goal.

Subsidy payments have sky-rocketed. Although they rise and fall in counterpoint to the markets, they increase especially after increased periods of "success" and "efficiency" on the farm. Simply put, increased production suppresses prices below the cost of production, and the government makes up the difference. One of the better examples came after a series of great plow-ups in the late 1970s spurred by Earl Butz, Nixon's controversial secretary of agriculture, who urged farmers to expand by plowing "fencerow to fencerow," or, in other words, to put all land into production. Farmers took the advice to heart, tearing out the windbreaks, shelter belts, and filter strips that had been meticulously planted (at the urging of earlier administrations) to repair the effects of the Dust Bowl's erosion and prevent its recurrence. Once again, farmers pushed across the borders of the arable world, plowing up land not fit for farming. And, once again, they overproduced. Prices fell, and through the 1980s direct farm subsidies rose to an average of ten billion dollars a year.

This led to a backlash of sorts, and in the late 1990s a Congress flushed with a messianic belief in the free market system passed what was then called the Freedom to Farm Act, a piece of legislation that was meant to phase out subsidies. The result? Payments that averaged ten billion dollars a year in the decade leading up to the mid-1990s tripled to almost thirty billion in 2001. That increase generally gets blamed on ever-falling commodity prices, and that is the direct cause. More important, though, farm legislation has become over the years progressively more generous to farmers. It seems counterintuitive that this could stand as a political reality when the number of farmers is decreasing. One would think the downward trend would translate into decreased political clout. A big part of the explanation of this anomaly lies in the capital-intensive nature of farming, which makes farmers the conduits, not the recipients, of subsidy. Just as industrialization allowed farming to conquer a new land base, it allowed farming to cross political sectors as well. Oil companies, tractor dealers, machinists, canners, migrant workers, gasohol plants, chemists, and gene jockeys are all a part of the farm lobby.

Eastern Montana is wheat country, encompassing an area called the Golden Triangle that is carpeted with wheatfields and fallow wheatfields, seemingly unbroken for hundreds of miles. It is on the ragged edge of arable land, meaning the soil is thin, dry, and highly prone to erosion. I once asked a county soil conservation official how much of Hill County, a countryside covered with wheatfields, is inappropriate for wheat farming. "All of it," he said. Walking the margins of this land, where the wheatfields run up against the drought-parched hills that were once short-grass prairie, gives a glimpse of why this landscape nevertheless persists. The hills, the worst soil, are dotted with small pocket wheatfields, unharvested in late fall. The wheat is short beyond dwarfing, seed heads shrunken and useless, planted on lands never meant to be planted. The farmer who planted them knew it. The strategy on these marginal fields is to take advantage of a year or two with a bit of extra moisture, plow up the fields, and establish a record of cultivation. That record qualifies the farmers for subsidies, even on the poorest land. The farmers know full well that these fields will fail eight years out of ten. That is, they will fail to produce wheat. They will, however, grow money every year.

Of course, this is the strategy on marginal lands, and not all lands are marginal. In the places that produce year after year, the new technologies produce abundant harvests, not only in America's mid-section, but increasingly, around the world—in the neo-Europes, but also in places like China, Mexico, and the Punjab region of India. Wheat, maize, and rice are chronically in surplus worldwide, largely because of the "success" of the Green Revolution. Yet we continue to grow more and more of these very crops, not because they are needed but because we know how to grow these crops—and these alone—abundantly, easily, and well. This is the situation that has caused the world to rely on three grain crops for more than two-thirds of its nutrition.

All of this grain flows into a stream that feeds either poor people or livestock. In the developing world, people take in most of their calories simply by consuming grain directly. Yet per capita grain consumption is much higher, as much as three times higher, in the rich parts of the world. Most grain consumption in places like the United States is indirect, however, through consumption of livestock fattened on grain. Increasingly, though, even animals can't consume all the surplus, which explains why midwestern corn farmers are lobbying intensely for gasohol made from corn, a way to pour their surplus production down the limitless gullet of American automobiles.

American farm experts no longer speak of food, but of "commodities." And with reason. The produce of farm fields is no longer a diverse flow of foods to tables so much as inputs into a series of factories. Livestock have left the farm and are now produced in what are called "confinement operations," beastly concentration camps where chickens, hogs, cattle, and turkeys are packed and fed a stream of grain. It's too expensive to recycle their manure to fields hundreds of miles away, so it becomes waste, a term unique to industrial thinking. Seventy-eight percent of American beef comes through these feedlots. The remaining grain, meanwhile, goes to processing plants. Most of our corn not fed to livestock is processed into corn syrups. All of this has conspired to render American agriculture an enormously narrow system. In the year 2000, 85 percent of the country's cropland was planted in four crops: corn, soybeans, wheat, and hay. Of the four, only wheat is even close to what we think of as food—something people eat directly—and even that must undergo processing to become flour. The corn is grain corn, not sweet corn, and must be processed. The same with soybeans. All of the rest goes to feed livestock or to food processing factories. The ramifications of this narrowing for the quality of our lives are the subject of a later chapter; the point here, though, is to understand how this system bumps up against the border, the limit of arable land, and its ecological devastation of a range of landscapes.

We may begin looking for agriculture's encroachment not in an Iowa cornfield but far out at sea, in the Gulf of Mexico. This is obviously not farmland, but it would be hard to find an ecosystem more radically changed than the gulf during the last generation. Despite at least a century's worth of industrial abuse from oil refineries and chemical plants dotting the Mississippi Delta, the gulf still managed to survive as a relatively intact ecosystem until a generation ago. It produced vast amounts of fish and shellfish, especially shrimp, that factored richly in the American diet. No more, at least not in the vast region of the gulf ecologists call the "Dead Zone."

The effects of modern industrial agriculture range from pesticide pollution to freshwater depletion, energy consumption, erosion, and salinization. We can, nevertheless, trace the Green Revolution's swath across the planet, especially in marine systems, by focusing on a single element—nitrogen. Nitrogen figures in problems as diverse as red tides, fish kills, marine mammal deaths, shellfish poisoning, loss of seagrass habitat, destruction of coral reefs, and acid rain. Beginning in about 1950, the use of nitrogen fertilizer ballooned from less than five million tons annually worldwide to about eighty million tons today, the result of employing chemical fertilizer to pump up seed heads on dwarfs.

This transformation in the world's supply of nitrogen has shown up heavily in terrestrial systems, largely in the competitive advantage given to nitrogen-limited species. This is what is causing, for instance, grassland to replace highland heath in Scotland. Fertilizers also readily convert to atmospheric nitrous oxide, a global warming gas. Most of the nitrogen, however, leaves farm fields with runoff, so the most apparent damage is to rivers, wetlands, estuaries, and seas, where it causes eutrophication, anoxia, and hypoxia, various types of oxygen depletion as a result of excess nitrogen. Dots of these water-borne problems pock the globe, wherever farming touches water, but the problem is most easily read in the Gulf of Mexico, which now bears a twenty-thousand-square-kilometer hypoxic Dead Zone. Fish and shrimp have disappeared from this area; 85 percent of the gulf's estuaries are affected. The nitrogen causing this all comes from the Mississippi River, which drains a vast region of the United States, but an Army Corps of Engineers study was quite specific about the source. Seventy percent of the Mississippi's nitrogen comes from a relatively small six-state area that is the heart of the nation's corn belt.

Nitrogen is only the best indicator of this flow. Loss of biodiversity, pesticide pollution, nitrogen pollution, soil depletion, erosion, siltation, eutrophication, desertification, salinization—all of these are part of the long list of problems that inevitably stem from agriculture. There has been some debate as to which of these phenomena is most threatening, that is, which ought to take priority in the search for solutions. Generally, though, issues of water quality are the most alarming. The United States Department of Agriculture conducted a survey of farmers and farm experts in 1994 that showed a consensus; respondents ranked water quality as the top problem, with soil erosion, sustainability, land conversion, and pesticide management just behind. The Environmental Protection Agency says simply that agriculture is the primary cause of water-quality problems nationwide.

The U.S. Geological Survey determined in 1994 that 71 percent of U.S. cropland is in watersheds where at least one pollutant's concentration exceeds standards.

Further, there is reason to believe, especially as a result of research completed after the USDA survey and only now settling into our consciousness, that we underappreciate the seriousness of the single problem of nitrogen pollution. It is becoming clear that human-produced nitrogen, largely from fertilizers, is altering the conditions of life on the planet. Humans now contribute more nitrogen to the global cycle than do all natural terrestrial sources. This scale threatens consequences equal to those caused by human disruption of the carbon cycle.

The concentration of row-cropping corn and soybeans in the upper Midwest, however, suggests that we not put too fine a point on ranking the problems. Corn and soybean production is the leading contributor to all of them. To a large degree, each problem is associated with the fact that water runs downhill across bare soil, taking with it the soil and the pesticides and excess fertilizer that it contains. Add to the equation the fact that soil must be bare of other living plants to support monocrop, annual agriculture, and you include the problem of loss of diversity.

This killing of the gulf raises the issue of resource allocation in addition to the environmental issues. The Dead Zone has already seriously damaged what was once a productive fishery, meaning a high-quality, low-cost source of protein is being sacrificed so that a low-quality, high-input, subsidized source of protein can blanket the Upper Midwest.

These problems are almost guaranteed to arise from the high concentration and volume of acreage devoted to corn and soybeans, but also because of the fact that corn farming accounts for 57 percent of all herbicides and 45 percent (by pound of active ingredient) of all insecticides applied on all U.S. crops. Soybeans account for another 19 percent of herbicides, though only about 1 percent of insecticides. Together, these two row crops are the leading contributor to soil erosion and the leading source of groundwater pollution from both pesticides and nitrates. Corn is the second leading user of irrigation water. The Mississippi and any other watershed that drains farmland also carries a load of pesticides, as well as soil eroded from farm fields.

The steppe lands around Asia's Aral Sea were so arid that they defied agriculture throughout most of recorded history and, accordingly, raised a defiant culture of pastoral nomads, the Mongol hordes. Plowing these lands made them no less arid, so their thirst needed to be slaked with irrigation. The Aral Sea's water had been used for growing crops for thousands of years, especially in the regions bordering the Amu Darya and Syr Darya, the sea's two major source rivers. Beginning in 1960, however, irrigation began to draw more water than the sea's attendant rivers could supply, precipitating one of the globe's more pronounced environmental catastrophes. Irrigation of crops, particularly cotton, in the Central Asian states, especially in what is now Uzbekistan and Kazakhstan, has shrunken the sea to less than half its former area and claimed about 75 percent of its former volume. The shrinkage has increased the sea's salinization and killed its fish. A body of water that was once a thriving and sustainable source of protein effectively became fishless by 1980.

Irrigation is as old as civilization, and through most of history twinned itself with rivers. The world's great valleys, such as the Nile, the Tigris and Euphrates, the Yellow, and the Yangtze, depended on periodic floods to replenish cropland with both water and fertile silt. The culture was integrated with the natural course of events. Nonetheless, even this low-tech form of irrigation required some organization to see that canals were built and maintained and to decide who had access to water. These peculiar requirements of irrigation gave rise to particularly hierarchical and rigid societies, noted by Karl Marx and labeled "hydraulic societies" by the German philosopher Karl Wittfogel in the early twentieth century. The environmental historian Donald Worster has extended Wittfogel's thinking into our time, especially in his book *Rivers of Empire: Water, Aridity, and the*

Growth of the American West. Worster argues that hydraulic societies have entered a new period he calls the Capitalist State Mode:

> Where water control is carried out comprehensively these days, it is by means of modern technology—electric pumps that can lift an entire river over a mountain range or mammoth concrete dams that create artificial lakes over a hundred miles long. The early hydraulic societies, organized along agrarian state lines, have all now disappeared, along with the apparatus they operated. In their place stand the new modern hydraulic societies, the most developed of them spiraling outward in the American West, and these societies express the reigning mind of the marketplace men, the technological wizards, and the ubiquitous state planners.

If anything, modern irrigation has spawned a culture even more rigid and hierarchical than before—the social cost of the technology. The environmental cost, however, is even more pronounced. In the United States, for instance, the entire Colorado River basin has been appropriated, mostly for irrigation, so that the Colorado no longer flows into Mexico's Sea of Cortez, triggering the death of that productive bay and its estuaries. The Oglalla Aquifer, which underlies about five states in the United States' southern plains, is nearly depleted. The Columbia River no longer supports salmon, partly because of irrigation. China sends tankers to southeastern Alaska to load up with nothing more than river water for drinking, so scarce has freshwater become in the oldest surviving irrigated civilization.

In the United States and worldwide, the land now farmed has simply appropriated the natural water flow—the lifeblood—of other lands. The rain that falls in mountains and deserts once fed streams, once fed habitat. Now those regions are farmed—not directly, but for their water. Irrigation now accounts for 70 percent of the freshwater used by humans. Again, this appropriation did not grow in a long, continuous curve from the beginning of agriculture but is a modern phenomenon. During the last forty years, the amount of irrigated acreage in the world doubled. The doomsayers predicted famine in the late 1960s, largely as a result of a swelling population's bumping up against the intractable limit posed by the planet's finite supply of arable land. We jumped that limit, but did so by spreading the footprint of farming to mine, sterilize, and dewater the rest of the land, not to mention estuaries, gulfs, rivers, lakes, and the atmosphere itself. We no longer grow crops just on land; we have plowed up the biosphere.

A VANGUARD OF FEUDALISM

The rush of stunning headlines in 2001 shoved many significant events from our collective memory, just as others were indelibly burned in. No doubt, the report from *The Economist* about a surplus of wheat in India fell into the category of the lost. Still, it held some jarring descriptions for those of us raised to think of India as a famine center:

> Rats and buffaloes in the Punjab, India's breadbasket, are in fine fettle. The rodents are feasting on millions of tons of wheat and rice stored in government warehouses (or, frequently, in the open air), the cattle on discarded potatoes. But no one else is happy. The government cannot afford the huge cost of buying and storing the grain coming from the farms in the Punjab and elsewhere in India, nor can the poor afford to buy it.

This ought to have been sobering news to a generation of food experts whose reason for being was to raise grain yields. This is one result of their good work. The fundamental shift in the basis of agriculture that we examined in the previous chapter spread many of its problems, while at the same time creating some new ones when overlain on conditions in the developing world. The Green Revolution has proved to be an ambiguous blessing.

There can be no doubt that the Green Revolution had benefits; it did indeed stave off famine, and India is in fact the best measure of its success. Nonetheless, we are left wondering how much of this success we can stand. This is a question in the United States, but technically, the United States was not the focus of the Green Revolution. American farmers have always grown commodities in surplus. Our country is almost unique in having never known famine. Even during the worst of the breadlines in the Great Depression, there was surplus grain. Furthermore, with the introduction of hybrid varieties of maize, dramatic yield increases in the United States preceded the Green Revolution. Nonetheless, the Green Revolution abroad not only paralleled developments in the United States but fed its burgeoning yields into the same global markets as did (and do) American farmers. The differences that demarcated the developed world from the undeveloped world at the outset of the twentieth century faded over the next few generations, at least down on the farm. Rats feeding on surplus grain in the Punjab are one measure of that change.

Indeed, given the history of the Green Revolution, the Punjab in particular and India in general are fine places to begin tracing its effects as they bear on communities and individuals. So, too, is Pakistan. Better still, the Yaqui Valley of Mexico, where Norman Borlaug began his work.

The region that holds the Yaqui Valley is most strongly associated in North American consciousness with places like Mazatlán, Cabo San Lucas, Acapulco, and Puerto Vallarta—tourist traps and colonies of gringo resorts. Among environmentalists, the region has a different claim to fame; the Sea of Cortez (the Gulf of California to many) is one of those sad seas of the world dying from dewatering, overfishing, and increased salinization. The biggest share of the blame for this goes to farming. The Colorado River is the Sea of Cortez's primary source of freshwater, flowing into its northern edge; before the Colorado River crosses the border, Americans are pumping it dry for irrigation and, increasingly, municipal water. Between the tourist drek in the south and the problems from the north, about halfway down the sea's eastern shore, lies the Yaqui Valley, a discrete little laboratory for measuring the aftershocks of the Green Revolution.

As you come to the edge of Ciudad Obregón, the Yaqui Valley's urban hub, you turn down Avenida Norman Borlaug, a main drag. Then you enter a compound that consists of a clutch of low sheds and greenhouses surrounded by squared-off test plots of wheat. It is a field station like any field station around the world, distinguished only by its seminal role in the history of agriculture. For this is Borlaug's original work station, now operated by CIMMYT, the Spanish acronym for the International Center for the Improvement of Maize and Wheat. CIMMYT is the eldest among what is now a series of similar institutions known collectively as the Consultative Group on International Agriculture Research, or simply the CG system. The Rockefeller Foundation seeded the network as a way of institutionalizing the Green Revolution and still foots a good share of the bills.

Inside the station, we are greeted by Ravi Singh, a plant pathologist, and Ivan Ortiz Monasterio, an agronomist. They represent the mix of international expertise one can expect to find at almost any CG system station around the world. The mix is a key part of the revolution, as important as the increase in yield per acre. As the system developed, the foundations involved consciously promoted the training of scientists from the developing world, usually at land grant colleges in the United States. Singh is Indian. Ortiz Monasterio, a Mexican, earned his Ph.D. in the United States. They are part of a second generation of scientists that has come into its own and now runs what is truly an international system.

The system's lineage, however, can be traced directly to Borlaug, who keeps an apartment at CIMMYT's headquarters near Mexico City. Most of the scientists have worked with him, and mention his Nobel Prize a few minutes into any discussion. Their reverence for him shows as Singh explains the chain of logic—and a pivotal decision—that rebuilt agriculture almost by happenstance.

Borlaug showed up in the Yaqui Valley in 1944 to work on a wheat disease called stem rust. At the time, the disease was raising hell with the wheat crop in the United States' breadbasket, and also with Mexico's wheat crop, which is grown in winter. Outbreaks would begin in Mexico and spread north, so Borlaug headed for the source and set up a breeding program to develop a variety resistant to the disease. That early breeding work involved a decision that was to alter the course of global agriculture.

Plant breeding is a tedious business. A breeder crosses varieties that exhibit desired traits, hoping by this roll of the dice to produce an offspring with a combined set of traits like disease resistance, shorter ripening time, higher grain quality, drought tolerance, and increased yield. All of the stars must align.

Getting progeny with the hoped-for traits is not, however, the end of the line. The traits must hold up in successive generations; typically, a project will go through a decade of breeding and growing to confirm that the breeder has come up with a winner. Mexico, however, offered Borlaug a unique opportunity. Its lowland desert in the Yaqui Valley would produce a crop in winter, while more temperate highlands would produce a summer crop. If he worked with varieties that would grow under both conditions, in both long- and short-day growing seasons, he could cut the breeding period in half by producing two generations a year. More significant in its implications, though, was his choice of varieties that would grow in such a wide range as his foundation stock. He would produce a global wheat, a one-size-fits-all seed that would travel around the world. From a seemingly practical decision to cut field time emerged a defining hallmark of Green Revolution agriculture.

Borlaug began tackling rust resistance, but soon decided to breed dwarf varieties already developed in Asia into his foundation stock. Gradually, the shortness of the plants became more important than their disease resistance. In 1960, Borlaug released his first semi-dwarf variety. By definition, dwarfs are more efficient plants, because they trade stem for grain. As we have seen, though, the exchange also results in a stronger architecture that allows farmers to pump up the grain yield with chemical fertilizers. The most important fertilizer in this process was nitrogen. Because the dwarf varieties' potential range was global, the extensive use of nitrogen spread with the seed. As a result, the nitrogen problem is global, too.

It is this problem that Singh and Ortiz Monasterio are now studying in the Yaqui Valley, which has once again provided an ideal setting for research. Unlike most of Mexico, the arid Yaqui River valley surrounding Obregón doesn't have an agricultural history stretching back to Aztec farming but was colonized only in modern times by large-scale commercial wheat farmers. The Yaqui is a leading edge of large-scale commercial agriculture in Mexico and, as such, a predictor of the shape of developing world farming. What we see today in the Yaqui is what the world will become, in the absence of a fundamental change.

The valley was a solid predictor of the nitrogen boom. Farmers in the Yaqui Valley make their counterparts in the American Midwest look like nitrogen misers. Some farmers apply about 250 kilograms per hectare of land, more than double the U.S. rate. National average rates vary widely around the world, from a low of 10 kg/hectare in sub-Saharan Africa to a high of 216 kg/hectare for a region in East Asia. The world average is 83 kg/hectare. Globally, the growth in the trend of fertilizer consumption is weighted to the developing world, largely because Green Revolution methods emphasized its use and spawned a series of government subsidies and investments. In 1960, the developing world accounted for 12 percent of all consumption; today that figure is 60 percent. The developing world has already overtaken the developed in its ability to replicate the Dead Zone.

The Yaqui Valley's connection to water also makes it a particularly apt case study. It is an isolated splotch of 225,000 hectares of irrigated agriculture surrounded by desert, by and large a single watershed formed by the Yaqui River. It is heavily watered by old-fashioned flood irrigation, creating a network of nitrogen flows that drain to the Sea of Cortez, a relatively discrete system, but one that embraces a vital web of estuaries, tidepools, mangroves, and marine life.

Yaqui made sense as a laboratory because the problem there was well-defined, the scale was graspable, and the threat was clear. Or so concluded Rosamond Naylor, an economist and senior fellow at Stanford University's Center for Environmental Science and Policy, and Pamela Matson, then an ecologist at the University of California at Berkeley. The two joined forces at a conference on global change in Aspen, Colorado, in 1992, when the discussion led to agriculture, and from there to nitrogen.

"We began thinking about this as a place that could go down the tubes," says Matson. "Anoxia [oxygen depletion as a result of nitrogen pollution] drives off anything that can swim, and it kills everything else."

Now Matson is at Stanford, and the Yaqui Valley project includes not only economists and ecologists but agronomists, engineers, geologists, hydrologists, climatologists, biogeochemists, political scientists, and geographers, all combining efforts to model and measure the interweave of social, biological, and physical flows of the valley. Based at Stanford, the project has grown to a $2-million-a-year endeavor to study intensively how world agriculture threatens biodiversity, and has made the Yaqui the most closely observed case study of agricultural nitrogen use in the world.

Not far into the project, Matson and Naylor teamed up with Ortiz Monasterio, the CIMMYT agronomist. The three began testing an idea.

Agricultural nitrogen, unlike most pollutants, is not a by-product of an industrial process. It is a purchased input. The nitrogen that finds its way into waterways as runoff is waste—an asset that got away. Field trials in the United States have shown that typically 50 percent, and sometimes as much as 70 percent, of the fertilizer applied simply dissipates. A good measure of this waste can be attributed to government policies and subsidies that make nitrogen too cheap to conserve.

The researchers put together trials to compare the practices of Yaqui farmers with simple alternatives that timed the application of nitrogen with stages of plant development and the irrigation cycle. The idea was to apply less fertilizer, but apply it exactly when the wheat would most readily absorb it, thereby minimizing runoff. The researchers concluded in a 1998 paper in the journal *Science* that these alternatives could cut fertilizer use to 180 kg/hectare, compared to the 250 typically applied by Yaqui farmers, thus greatly reducing nitrogen flows to the atmosphere and to the water without decreasing crop yield. The bottom line was a 12 to 17 percent increase in a farmer's net profit. To a conservationist, such a result is not just a number but a tool, the elusive win-win that motivates change.

This sort of gain is of particular concern to those who work in the developing world. The United States has a long history of regulating pollutants but still is hard pressed to regulate nitrogen flow from farms, which is diffuse and difficult to control. It's worse in the developing world, where lax, ineffective government on one hand, and the ubiquity of hunger and poverty on the other, make governments even less willing to experiment with policies that might jeopardize grain yields. Enter the unseen hand of the market. The investigators quickly found out they had a potential market solution to an environmental problem, and set out to test it.

In a multipronged effort to spread the glad tidings, they used a series of contacts with local elites, set up farmer field schools, planted demonstration plots, and pamphleteered. The interdisciplinary program took into consideration the fact that farmers ranged from major commercial operations like those in the United States to twenty-acre dirt farms and dirt-poor cooperatives called *ejidos*, part of the peasant land-distribution system set up by the Mexican Revolution. The message was tailored in content and delivery to reach each stratum in the complex social structure.

In the spring of 2001, Naylor, Matson, Ortiz Monasterio, and I visited a cross-section. The researchers went into the field expecting a payoff for the years of demonstrating their idea, especially as they had gotten an unexpected boost from a recent turn in global markets. Fertilizer was already the costliest element of a farmer's annual investment when they had done their initial analysis in 1998, but in the interim, there had been a sudden 50 percent surge in its cost. Conservationists in the United States have suggested that even a relatively modest tax on nitrogen could help stem the flow, so the Yaqui Valley was offering a real-life test in the extreme.

Juan Dorame has a simple little farm smack in the middle of the Yaqui. We walk his neat field of raised durum wheat beds as he details for us with some pride his bounty yields, which are accomplished with less irrigation water (read: less runoff) and on average about 20 percent less nitrogen than the norm. Oritz Monasterio uses some of Dorame's fields for trials, making for a clear demonstration to his neighbors. Dorame tells us through a translator how he has invited his neighbors to come to his fields, to literally count every stalk in a square meter and to weigh every seed head to check his yield. Their response? They don't believe him, and they change not a thing on their own farms.

"It's hard to change these people," he says. "They think it is not necessary to change what they have done for twenty years."

Jorge Castro runs his family's farm of 250 hectares, big by the valley's standards. An innovator, he has allied himself with a couple of neighbors to found a research organization and is experimenting with marketing his own brand of seed maize. He grows safflower, soybeans, potatoes, and, of course, durum wheat. He practices conservation tillage, which reduces runoff. He and his banker have begun thinking in terms of maximum economic yield, which is to say he will accept some decline in his yields as long as the reduced costs maximize profit. This sort of thinking and Ortiz Monasterio's demonstrations have caused him to cut nitrogen rates from 250 to 180 kilos per hectare.

So here is a bellwether farmer, one whose practices can be counted on to provide the example that will sway neighbors. We ask Castro about this. He shakes his head for a moment and fiddles with his cell phone. The agronomist's tool for spreading the word is to hold field days, when neighboring farmers can see for themselves the profitability of progressive practices.

"We have held many, many, many field days," he says through a translator. "There has been no impact. This culture is deep-rooted. It is hard to change."

Enrique Orozco Parra is not just a farmer with five hundred tilled acres under his control. He is also president of the Patronato, an organization responsible for research, innovation, and information to which every farmer in the valley belongs. A conversation in his office runs from water use to subsidies, global markets, and costs of capital.

He has a solid command of it all. And, yes, he has heard the business about reducing fertilizer use, has seen the test plots, and, yes, the 50 percent price increase has been a terrible burden on profit.

So how has he responded on his own land? No change. In fact, sometimes he lays down as much as 300 kilos of nitrogen per hectare, an amount off the global scale. In his excellent English, he tells us that nitrogen is cheap insurance.

Anecdotal, all, so after our visits, Ortiz Monasterio followed up with a survey of farmers. After five years of work, no change. The Sea of Cortez is bearing the same burden as always—in fact, as we shall see, probably more. Nor should it really surprise anyone. Expecting farmers to respond to market signals now is a bit like expecting an alcoholic to order the herbal tea at an open bar. This is the legacy of subsidy. Governments, including Mexico's, got in the business of making nitrogen cheap, and farmers lapped it up, but it created a welfare state. Emblematic of that state is a deep-seated irrationality that ignores cause and effect. Farmers worldwide operate in an economic never-never land where governments escalate subsidies and other protectionist measures as a sort of arms race, a system that has taken on a logic of its own. No single set of market solutions will turn that system around.

It has long been argued that intensive agriculture is a sort of global sacrifice zone that, in the end, benefits biodiversity. That is, surrendering land to intensive and efficient agriculture frees surrounding lands from extensive agriculture uses like grazing, and effectively saves habitat. The researchers didn't go into the Yaqui expecting to test that notion, but Naylor says that test is happening. The scientists from a range of disciplines who have been scrutinizing the Yaqui for these years can't help but notice that change happens.

Monoculture is not just a problem for nature; the lack of diversity threatens farmers as well, especially when pacts like the North American Free Trade Agreement plunge them into global markets, and especially when an entire valley relies on a single crop like wheat for most of its income. The farmers of the Yaqui understand this better than anyone, so they are using that access to markets as well as capital generated by past development to diversify. Expect this to happen in the rest of the developing world as well. Is this a good thing? It can be, when, for instance, farmers diversify row crops into something like chickpeas, as many have. Because chickpeas are a legume, they fix free nitrogen from the air, reducing the need for synthetic nitrogen applications—at least on paper. In interviews, though, farmers said they regarded this as simply a shift to another crop and were not taking advantage of well-planned rotations to cut back on nitrogen applications. Nor have agronomists in the region developed such rotation systems. And Ortiz Monasterio says experiments have demonstrated that legumes actually decrease their rates of fixing free nitrogen from the atmosphere when synthetic nitrogen fertilizer is in the soil, so in the Yaqui's environment, a potential solution fails.

More significant, however, is the valley's growing diversification into livestock and aquaculture. The former is attractive to farmers because it provides a huge sink for surplus crops, a sort of value-added operation that turns superfluous grain into meat readily snatched up by global markets. There is a lesson here that is significant for all of the developing world. Ostensibly, the Green Revolution boosted yields to feed the poor, but as its successes leave the globe awash in cheap grain, the tendency is for that grain to go to livestock that feed the world's rich. Further, grain-fed livestock operations act like nitrogen factories.

Miguel Olea walks us through his 6,000-sow hog farm, a typical confinement operation of the sort that has inflamed political debate in the American Midwest. The brood animals are penned in spaces no larger than their Harley Davidson–sized bodies. The hogs ingest a constant stream of feed at one end and send a constant stream of nitrogen-laden manure out the other. Because of inefficiencies in the way they process food, a typical hog will produce about triple the waste that a human will in a day, with the result that a hog farm of 6,000 sows and their young produces as much untreated sewage as a sizable city. Olea's sows produce about 100,000 young hogs for slaughter each year, feeding a globally stratified market in which the expensive cuts are shipped to Japan and the cheap cuts stay in Mexico.

A couple of generations ago, the phrase "living high on the hog" still had a direct meaning. In the United States, even well into the second half of the twentieth century, a few hogs were a part of every farm, providing a sort of living recycling system, especially on dairy farms. Waste milk, skim milk, waste grain, kitchen scraps, and garden leavings all went to the hogs, which dutifully converted them to protein, and then valuable manure. In the fall, the farmers butchered the hogs, and the family would recover the protein and fat. For a few days, the family would eat the expensive cuts, chops, which come from the soft muscle along the backbone, high on the hog. Later they would eat the rest.

Slowly, this system broke down as hog production industrialized, concentrating literally hundreds of thousands of hogs on a single farm. This evolution was somewhat gradual. At first, a few hogs went to a local butcher, who sold them, a few cuts at a time, to the community. Those who could afford to do so bought the more expensive cuts, living high on the hog all the time, and the less well off got the rest, a sort of hog hierarchy. My father tells of classmates in his rural Michigan community eating lard sandwiches for lunch every day during the Depression. Now, however, that stratification is global. It is easy to believe in progress when you export the lard, and the lard-sandwich eaters are conveniently out of sight, half a hemisphere away.

Out back of Olea's massive steel barns, a simple slit of a drainage ditch carries a steady stream of manure slurry, where wading egrets look for an early cut of the nutrients. In the best case, this nitrogen-rich stream flows, as Olea's does, to fields, in place of synthetic nitrogen. Even there it runs off in storms and after flood irrigation, to join the stream bound for the estuary. Already there are about thirty thousand sows in the district around Ciudad Obregón, 103,000 in the state of Sonora.

Along the drainage canals that parallel the valley's straight roads heading for Tobari Bay and the Sea of Cortez there are no hogs in sight, but everywhere there are billboards advertising bank financing for shrimp farmers. This is the more significant track of diversification, driven by a huge demand for shrimp in Europe, the United States, and Japan. Worldwide, shrimp farming has become an explosive threat to biodiversity, largely because it operates at such a high trophic level. Shrimp eat protein, fishmeal vacuumed from marine sea webs by factory trawlers. Moreover, in Southeast and South Asia, Latin America, and, increasingly, Africa, these farms are located next to estuaries, usually in mangroves, which are locally regarded as scrub land but are in fact the biological key to the productivity of tropical estuaries. The farms concentrate nutrients and disease, providing an effluent enriched not only in nitrogen but also phosphorus, antibiotics (routinely fed to the shrimp), and fungal diseases. The area of Mexico surrounding the Sea of Cortez now holds at least 26,000 hectares of shrimp ponds, which send about 3,000 tons of nitrogen to their waste streams annually. Potentially, these numbers could multiply by a factor of ten.

The term for this artificial feeding of confined fish to produce food is "aquaculture," but do not imagine that we have left our subject and gone to sea. Aquaculture is an extension of agriculture, an attempt to shoehorn the marine world into the system of farming. It represents yet another dramatic and largely unnoticed trespass of agriculture beyond the limits of arable land.

Human reliance on fish protein predates farming. As we have seen, fishing played a key role in allowing the sedentism ten thousand years ago that led to the domestication of plants. Aquaculture itself is ancient, especially in Asia, where carp have been farmed for millennia; the book *Fish Culture Classics* was in its first printing in 460 B.C. By and large, though, most of the fish people have eaten have been wild, and today's fishermen are our best remaining example of hunter-gatherers. At least hypothetically, fishing is sustainable, in that it relies on intact ecosystems to produce a surplus of fish, a harvest that can go on forever. Of course, it hasn't. The world's oceans have been exploited as heavily as the rest of the planet, especially as the technological means for doing so have increased. Today's fishermen rely on multimillion-dollar vessels, computers, satellite-assisted geographic positioning systems, radar, and sonar to find fish. Thanks to these methods and population pressure, fish stocks are in peril worldwide.

Fish farming advocates argue that farmed fish take the pressure off wild species. In fact, farming has filled a void. Between 1986 and 1996, the amount of fish produced by farming more than doubled, with the growth concentrated in two species, salmon and shrimp. Well over half the salmon eaten worldwide comes from farmed fish.

Has this helped wild salmon, a species clearly on the ropes in much of its habitat range? Both salmon and shrimp are carnivores, that is, protein eaters. Further, they are evolved to eat only fish protein. In a paper in the journal *Nature*, Naylor, along with a team of biologists, concluded that each kilo of salmon raised by aquaculture requires 3.16 kilos of wild fish for meal and oil. A kilo of shrimp requires 2.81 kilos. As with hogs, there is an issue of equity here. Fish farmers argue that they are laboring to feed a hungry world, but overwhelmingly, farmed fish and shrimp go to wealthy tables in Europe, Japan, and the United States. Increasingly, living high on the hog means living high on the food chain.

The protein that supports this practice comes from what are called "trash" fish, such as sardines and herring, as well as fish waste. Many of these species, however, are not trash at all, but a key link in the ocean's food chain and an important part of local fisheries and diets in the developing world. Left to their own devices, wild fish (especially salmon) eat the fish that factory trawlers are now scooping up and grinding into fish meal to feed to farmed fish. Absent the trawlers, local fishers in skiffs or pirogues would catch a few anchovies to feed local protein-starved communities. Instead, this protein is sucked up, reduced by a factor of three, and shipped to Red Lobsters across suburban America. The three pounds of protein that raise a pound of salmon for rich tables come from the protein budget of the poor.

A walk through a thousand-hectare shrimp farm, one of several now winding around the valley's border with the sea, reveals a flat landscape completely stripped of vegetation and diced into ten-hectare "tanks," which are simply rectangular ponds held in by bulldozed berms. Each hectare will produce more than two tons of shrimp. An intake canal pulls salt water from the sea, two kilometers away. Eight 250-horsepower diesel pumps suck salt water to the ponds. A discharge canal sends effluent back at the same rate.

The managers of this operation, a cooperative that includes eleven local *ejidos*, make a point of telling us they have carefully situated the farm inside the ring of mangroves. This is true, but as we walk along the outlet canal—flush with juvenile heron, egrets, kingfishers, and pelicans—we also notice that the canal cuts straight across a tidepool, effectively damming tidal flows from a mangrove area easily the size of the shrimp farm. Along this stretch, the mangrove's trees are dead. The canal discharges straight into a mudflat that is shiny with algal slime at low tide. We can hear a cacophony raised by shorebirds and migratory waterfowl that depend on this key section of the Pacific flyway, but in the distance. The canal has washed old tires downstream with the rest, and they stand upended in the mudflats, planted like tombstones.

Paredoncito is a fishing village just coming awake on a sunny morning in January. Kids head toward school while women sweep the dirt streets and chickens scratch and peck. We stop and speak with Bernal Guadalupe, a fisherman in this village since 1970. He catches the estuary's native shrimp, crab, and fish and sells them, on a good day, for enough to buy a day's worth of gas for his boat and food for his house. A year before our visit, Paredoncito fomented a significant event, and Guadalupe wants to show us the scene, so he climbs into our beat-up Suburban and we bounce the back roads for a couple of kilometers to a broad, flat stretch of bermed shrimp ponds. This was to be a shrimp farm, but it lies dry and abandoned. Guadalupe and his wife, Marialena Garcia, take us to the discharge canal to show us why. They describe the day when all the people in the village and several neighboring fishing villages marched to the canal and filled it in, digging with shovels and their hands.

He tells us the villagers had heard that the shrimp farm would harm the native fishery. He's right; shrimp farming kills wild shrimp. So they stopped the farm. Such matters are taken seriously in the Yaqui, where there has already been one murder stemming from confrontations between fishers and shrimp farmers. The protest brought police. Then Mexico's Secretary of State came. The villages had tried to stop the farm through more formal action, but that had failed to get the government's attention. The dig-in did, though, and the government pulled the farm's permit.

If we return to what it was that Henry A. Wallace intended, what the Rockefeller Foundation and the Green Revolutionaries intended for the Yaqui Valley, it would have been something like progress. Progress—a belief that we can make things better, that as time passes we and our technology improve lives—is an article of faith that organizes American society. Set this against a conversation we have at Paredoncito with one of the *ejidatarios*, a fisherman. We ask him how he makes ends meet now that fishing is bad, and he says he gets work now and again on a neighboring commercial farm, only he does not refer to his employer, a rich man, as a farmer. He tells us he works for the *patrón*, a politically charged term if ever there was one. It is a term of feudalism, a way of life the Mexican Revolution was supposed to erase almost a century ago.

I have come to think of agriculture not as farming, but as a dangerous and consuming beast of a social system. I think of an insect I once heard about that plagued a particular crop, chickpeas, until evolution finally taught the plant to secrete a protein that killed the pest. Then the bug evolved to convert that same lethal protein to its own use.

Progressives like Henry Wallace and Norman Borlaug went to work to help poor people. Wallace's hybrid corn did indeed do that for a time, making life better for a generation of farm folks. But in the end, it only enlarged the pile of surplus grain, which the system evolved to digest for its own purposes. The *patrones* of the world, the men made increasingly wealthy as the lot of small farmers has deteriorated, are testimony to the power of that system to sap progressive energy. Sixty years after it started, the foundations are still in the business of promoting agriculture as a cure for poverty.

People can and do thrive without cereal, or grains of any sort. Personally, I try to, although it's difficult to steer clear of its ubiquity. When I am in complete control of my diet, not traveling, choosing and cooking all of my own food, I get my carbohydrates from fruits and vegetables. The less cereal I eat, the better I feel, a reflection of my body's genetic legacy set during a couple of million formative years as a hunter-gatherer. Yet we as humans passed that point ten thousand years ago. There can be no doubt that a certain amount of cereal does have value—staff of life and so on. The pertinent question is not whether cereal has value, but, as an economist would put it: What is the value of more cereal? I offer the persistent need for farm subsidies worldwide (overwhelmingly for corn and wheat) as the market's expression of the value of more cereal. We can argue about foreign competition and preserving family farms from farm bill to farm bill, but the very existence of the subsidy means the market values that extra bushel of corn at less than what it costs to produce it. The endurance of those subsidies, year after year, through Republican and Democratic administrations alike, is the prima facie evidence that farming has evolved its own inertia, independent of human needs. The piles of surplus grain amid the poverty of the Punjab make the same case. Humanity's role is not to shape agriculture to its needs; rather, it is humanity's job to figure out how to pay for and dispose of all of that grain agriculture chooses to grow.

We live in an era of hyper-agriculture, and, as one would expect, the result is an exacerbation of agriculture's effects, especially a widening of caste, the gap between wealth and poverty. The gap exists within every agricultural society, but increasingly it has taken on a geographical overlay. In this sense, we translate the terms "rich" and "poor" into "developed" and "undeveloped" or "underdeveloped" or "developing" countries. Simply put, in the developed world, people derive about 31 percent of their calories directly from rice, maize, and wheat. In the developing world, that total is 56 percent (again directly): 18 percent from wheat, 27 percent from rice, 7 percent from maize, and 4 percent from other cereals. Yet even this equation masks the great divide, because the developing countries all have significant populations of wealthy people, and the wealthy don't eat grain gruels; meaning these percentages are concentrated on the poor. About half of humanity, three billion people, live on less than two dollars a day. Close to half of these, one-fifth of humanity, live in absolute poverty, on less than a dollar a day. Most of these poorest people are simply undernourished; that is, they don't have enough food. Most of the larger group, half of all people, are malnourished, in that they, like American livestock, eat mostly grain (which contains little protein, little fat, and few vitamins and minerals) washed down with filthy water. This is how grain agriculture invades bodies, just as it invaded streams.

TO SEE THE WIZARD

In deference to the oligarch who built this rail line, Amtrak calls the train the *Empire Builder*, but nothing eastbound passengers see in the first few hours of high plains Montana looks even vaguely imperial. This is the extreme western edge of a farm country unimaginably vast, unpeopled, and monotonous. As I ride this train east toward Cutbank, Montana, forlorn hills of grass, still snowbound in mid-May of 1996, give way to flat land tiled alternately with fallow strips and stubble strips of wheat, and so begins a monoculture of wheat and wheat farmers that stretches almost unbroken seven, eight hundred miles east to Minnesota, and, in variations, south through Texas.

Farther east, the land shades to a monoculture of corn that fingers still farther east, into the Ohio River valley. To see its sweep is to believe that it has always been so, but it hasn't, that it is a state of nature, but it isn't. This false face on the land has emerged only during the past generation. In truth, it is the face of empire, but some two hundred–odd years of agrarian myth and a few more years of being barraged by Archer Daniels Midland's non-commercials on National Public Radio and *The NewsHour with Jim Lehrer* have blinded us to what we are seeing.

In any event, few Americans have really laid eyes on this place. Most of us register it from a comfortable, pressurized distance, as "flyover country," that empty buffer between the nation's bicoastal poles of power, the clean geometric patchwork that looks like order and well-being from thirty-five thousand feet. There is a truer and more accessible image imprinted on every cereal box, Coke can, and candy wrapper, and echoed in those ubiquitous advertisements, in which an abstract of the landscape bleeds through the type: "You won't see our name on any of these labels. But ADM is on every one. . . . Check the labels of the food you're buying. Chances are you'll see at least one ADM ingredient, such as citric acid, lactic acid, high fructose corn syrup, sorbitol, lecithin, xanthan gum, wheat gluten, soy protein, and vitamin C, to name a few."

Any idea what this stuff is? It's us. It's the mass in our guts. It's our heartland. It's the molecules of our people and the juice of our politics, especially if you pay attention to the corn syrup.

I'm entering the story at one end, on a set of train tracks on the high plains, and you're at the other, entering through a cereal box, or a fifteen-second feel-good spot on *Meet the Press*. Between us lies Archer Daniels Midland, conduit of food and images of food, with fifteen thousand owned railcars, two thousand barges, a hundred oceangoing vessels, and a leased network of five million trucks and five hundred thousand railcars moving wheat, corn, and soybeans to 269 processing plants. Woven amid the rails and pipes is an integral web of effectively owned and leased politicians and news organizations.

Archer Daniels is not alone in this picture; it is just the standard-bearer. Still, the list of food processors is not terribly long. Five companies—Cargill, Incorporated; Continental Grain Co. (recently renamed ContiGroup Companies); Louis Dreyfus; Andre & Cie; and the Bunge & Born Group—control about 75 percent of the corn market. Four companies—Archer Daniels; Cargill; Bunge; and Continental Grain—control about 80 percent of soybean processing, both in the United States and globally. A single co-op, Ag Processing Inc., accounts for another 5 to 10 percent in the United States.

According to Thomas Jefferson's vision, this heartland was to be the anchor of national harmony, the centerpiece of a mythical agrarian identity antithetical to bicoastal poles of power. Jefferson imagined a democracy supported on the shoulders of yeomen settled as equals in the West, saw these open places being filled with good country people. Lincoln pursued this vision with railroad land grants and the Homestead Act, which were designed to give us independence, settlement, and stability. Instead they gave us instability, depopulation, oligarchy, and poverty. All that I am seeing—the landscape from the train window, the rush and blur that ties it all together, the cereal boxes—is not the whole of farm country, but it is a big enough slice to make me consider why the heartland is anything but, why the region that was meant to build democracy is, all the way from here south into Texas and east to the Upper Midwest, a beaten place.

I have an idea that Jefferson was not altogether wrong, if only in that we can see the truth of his vision in its negative. Our failure in the plains is the measure of the failure of our nation as a whole, a notion I will take with me as I travel backward on the same railroad that carried Manifest Destiny forward, then south to Decatur, Illinois. Farming is a pyramid. At the pinnacle of that pyramid stands ADM, the nation's largest buyer of grain, and at the pinnacle of ADM stands its flamboyant CEO, the empire builder Dwayne Andreas. He's the wizard we're off to see.

Sarah Vogel once got press because she was a lawyer for Ralph Clark, the right-wing nutcase who called himself a Freeman and staged an armed standoff with the FBI near Jordan, Montana. When I spoke with her, though, she didn't want to talk about the Freemen; she spoke of North Dakota's growing sweeter carrots than California. There is a connection nevertheless. Vogel is now North Dakota's agriculture commissioner, but during the farm crisis of the 1980s she lawyered in the cause of helping farmers fend off federal foreclosures. Like many ranchers, Clark had joined the government-sponsored binge of easy money that sucked agrarians nationwide into overexpansion. Of course the bust came, and the antigovernment activist, like most of his ruggedly independent neighbors, tapped the dole. Over ten years, he pulled in $676,082 in federal farm subsidies for his 960-acre ranch, but the money did not prevent foreclosure.

Vogel's father was a U.S. attorney during the fifties and sixties in the same stretch of plains. During his tenure, he handled three fore-closures on farms. During his daughter's first month as ag commis-sioner in 1989, the feds alone pressed 2,400 foreclosure cases in roughly the same territory. This has come to be known as the farm crisis, but in reality, it was but a single crisis in a long series, a brief uptick in a hundred-year trend in American agriculture. Throughout the period, boom-and-bust cycles broke farmers, and the land ac-creted in the survivors' hands. The survivors' new power helped them retool a federal system created to support family farms into one favoring the few, a trend that actually predates the crisis.

Coincident with this was depopulation; the nation's midsection has been unsettling in one long, unbroken curve. Population in the region peaked about the turn of the century. Nebraska now has as many as ten thousand deserted farmhouses. Kansas has six thousand ghost towns. Ninety percent of 450 rural plains counties lost popula-tion in the decade covered by the 2000 census, as they have in virtu-ally every decade since the 1920s. The people who are left are poor. About 16 percent of residents live below the poverty line in Mon-tana, Oklahoma, and Washington, D.C. Montana's people live in house trailers at more than double the national rate, a fact recorded in the ramshackle aluminum ring around every railroad town. We are at once the nation's breadbasket and its basket case.

This trend may be a century old, but something critical shifted during the 1980s, something that explains why Vogel speaks first of carrots and why so many of the elevators full of wheat now bear the logo ADM. Andreas and the rest of industrial agriculture (but espe-cially Andreas) advocate three things: exporting American agricul-tural products, keeping farmers on farms, and conserving topsoil. The 1980s farm crisis was precipitated by the first of these, a goal pursued at the expense of the other two.

From 1972 to 1981, American ag exports rose from $8 billion to $44 billion, mostly in corn and wheat. Much of this phenomenon can be traced to specific events, such as the USDA granting the Soviet Union $700 million in export credits in 1972, money the Soviets used to buy up a quarter of the U.S. wheat crop, which had the effect of nearly quadrupling the price of wheat. Andreas was then, and still is, a strong advocate of offering export credits, specifically to the Soviet Union and its successors, probably because the credits funnel straight back to ADM, which sells, stores, and ships the grain. Andreas personally claims credit for convincing Richard Nixon to sell wheat to China. The USDA now refuses to disclose how much ADM has made on the Russian deal over the last twenty-five years, but clearly the effects have remade the American landscape.

As the export boom took off, farmers borrowed heavily in order to specialize in corn and wheat, abandoning all other crops, and, with them, the time-honored practice of crop rotation. Within a decade, the export boom had produced record farm failures and soil erosion of Dust Bowl proportions. The boom caused overproduction; prices fell; interest rates rose; and international markets collapsed. Between 1970 and 1992, a million farms—36 percent of all American farms—ceased to exist. "Farmers are the most indispensable people on the planet," crows an ADM ad. Yet coincident with the rise of this corporation's fortunes has been an unprecedented decline in the number of American farmers.

Along the rail line, North Dakota is such an implacable sea of wheat as to suggest permanence, but these fields were once host to a diverse array of crops. The export boom hardened agriculture's monomania and the government bailout following the farm crisis tempered that edge. The government was hooked on using corn and wheat to balance trade, and structured subsidies to encourage those two commodities at the expense of almost all other crops.

Meanwhile, at the height of the farm crisis, the government abolished ceilings for key subsidized loan programs. Suddenly, there was a pool of bankrupt farmers, cheap foreclosed land and machinery, and subsidized capital—plus a government willing to guarantee income available to those willing to farm big. Today the richest 2 percent of all farmers—2 percent of 1.6 percent of the nation's population—account for 35 percent of total farm sales. At the same time, they receive 27 percent of federal subsidies. As many as 76 percent of the farms in some Colorado counties would lose money were it not for subsidies. Montana would have no net farm income—

zero—without subsidies. Half of the total federal income taxes paid by all North Dakota residents returns to the state as farm subsidies, yet only 9.5 percent of the residents of the state, the nation's second most rural, live on farms. This is a welfare state.

The predominant system of farming bolstered by all of this is accurately named industrial agriculture. It is capital-intensive, not labor-intensive, which largely explains the region's depopulation. Industrial agriculture considers the countryside as a factory. Capital buys machinery and "inputs" like fertilizer and pesticides. Farmers are, more appropriately, regarded as conduits, not recipients, of federal money, to the point that chemical suppliers, machinery manufacturers, and bankers have become the staunchest backers of the farm subsidy program. The output is corn and wheat, now in surplus, and in surplus virtually every year since the plains first saw a plow, in part because the subsidy system has encouraged surplus. These crops cannot be sold like a carrot can, directly to a consumer, but must be sold to processors like ADM. To a buyer, especially a megabuyer, surplus means cheap.

In 1878, when John Daniels was crushing flaxseed to produce linseed oil, processing was a fringe activity to farming. It still was in 1903, when a fellow linseed crusher, George Archer, joined the company. As it was in 1923, when the partners bought out a third linseed crusher, Midland Linseed Products. In the 1930s, after fifty-five years as a linseed company, ADM started milling flour, then processing soybeans, and finally expanded with the U.S. economy into overseas production in the 1950s. Dwayne Andreas, formerly an executive for another agribusiness giant, Cargill, took over in 1966 and held the reins until his retirement from the board in 2001.

When Andreas assumed control, ADM began a rapid expansion, from simply dealing grain to a broad range of food processing. This mirrored developments on the farm. As ADM diversified, farmers specialized, the better to feed ADM. Singlehandedly, the company buys about 12 percent of the nation's corn crop. It can store more grain than any other company in the world, and is the nation's largest miller of wet corn, processor of soybeans, and sheller of peanuts, and the second-largest flour miller. Out the other end of its pipe come flours, an array of vegetable oils, soy burgers, corn sweeteners, sugar, feed and feed additives, vitamins, vodka, and a seemingly endless list of additives for processed foods. To quote a business analysts' profile: "The agri-giant continues to grow by creating new products from, and new products for, its core corn, grain, and vegetable processing business."

ADM doesn't deal in food, it deals in commodities; thus it is wholly dependent on the system of federal subsidies that has converted American agriculture to one big commodity factory. In 1950, farmers grossed forty-one cents of every consumer's food dollar; that figure fell to twenty-one cents by 1994, simply because we eat more processed commodities. The 1994 figure of twenty-one cents, however, is an average. Even today, a farmer gets fifty-eight cents of a consumer's dollar spent on eggs, because an egg is food. A chicken makes it; a farmer puts it in a carton and sells it. When a corn farmer sells his crop to ADM, he gets four cents of the consumer dollar spent on corn syrup.

This observation is at the heart of an extraordinarily angry book, Victor Davis Hanson's *Fields Without Dreams*. Hanson, a failed raisin farmer in California's Central Valley, describes a mansion being built by a raisin buyer as all the farms around him are failing. The buyer "admitted in a rare moment of candor that you can make money in agriculture if you don't grow food."

A farm scholar once asked an agribusiness executive when his corporation would simply take over the farms. The exec said that it would be dumb for the corporation to do so, in that it is not free to exploit its employees to the degree that farmers are willing to exploit themselves.

Vogel talks about carrots because she is in North Dakota, the current leader in developing a cooperative marketing system for farmers. The state is alone in consciously trying to shake off monoculture through diversification and control of marketing. The largest of the new co-ops, Northern Plains Premium Beef, has 2,700 member ranchers in five states and two Canadian provinces. It's trying to circumvent the four beef packers that buy 85 percent of the state's beef. Wheat farmers have banded together to build a pasta factory at Carrington that employs two hundred people. There is an oil seed cooperative, an organic marketing co-op, and a co-op for bison meat.

These groups are among those trying to reintroduce a distinction between commodities and food, the latter being something people eat, like a carrot, and the former being something that a manufacturer like ADM buys and processes into products resembling food. The distinction between food and commodity says much about what we eat, but bears equally on the integrity of a farmer's work.

"When you're raising wheat and dumping it at an elevator, what difference does it make?" says John Gardner, who runs ag research at North Dakota State University. "But if you're raising pasta, it's food. You eat it. And all of a sudden it becomes a different thing. How you raise it, fertilize it, what pesticides you use—all of a sudden it becomes a whole different thing when it's food and not a commodity."

The distinction also dictates separate paths through the market, which is why Vogel worries about whether her state's carrots are sweet. The farmer puts the value in food, and its inherent quality recommends it to consumers. Commodities, however, leave the farm as an amorphous lump of grain. Processors make them into something like food, and packaging and advertising convince us it has value.

Corn, the nation's biggest crop, is mostly absent on the high plains; it can grow only to the east and south, where soils are thicker and summers hotter. But at Fargo, North Dakota's largest city and eastern edge, I leave the train for a day and drive a highway threaded through the absurdly black soil of corn's northwest frontier. On the radio, one of NPR's non-commercials lauds "Archer Daniels Midland, producers of lysine, an amino acid which promotes the growth of poultry and swine." (Based on an FBI investigation in the 1990s, ADM was convicted of fixing the price of lysine.) Beside the road, a billboard for Dakota Dinosaur Museum says: "You're on the Road to Extinction."

In fact, I am on the road to the three-thousand-acre organic farm Fred Kirschenmann has run since 1976. His parents settled the land in the 1930s, and his father was among the first farmers in the fifties to begin specializing and pouring on chemicals. In the late sixties, his father began to see something very wrong with the health of his land. Eight years later, after Fred finished his Ph.D. in religion, started teaching, and got ordained, he found himself "fed up with urban life." He returned to the farm, ready to try organic agriculture. His father said fine, do whatever works, and work it has.

"Not to be hokey, but I feel called by what I am doing," Fred says.

I bring up Clark and the Freemen. Kirschenmann mentions Gordon Kahl and Posse Comitatus, another homegrown band of agri-terrorists spawned by farm failures. In 1983, Kahl killed two U.S. marshalls at nearby Medina, then another lawman before he was killed after a cross-country chase. Good yeomen farmers hid him from the feds throughout. Posse Comitatus, Freemen, John Birch, militias, Aryans, Patriots—the plains seem to spawn a litany of such stories.

"We're dealing with a much bigger issue here: the extent to which we still have a democracy," he says.

Kirschenmann's neighbors grow wheat and (the lucky ones) corn and soybeans. Period. Kirschenmann grows oats, wheat, rye, barley, millet, buckwheat, alfalfa, flax, sweet clover, lentils, and sunflowers. He raises beef cattle, a practice virtually all of his neighbors have abandoned. There's not a lot of money in cattle, but they eat mostly his crop waste, and they produce manure. He sells them at an organic-market premium, as he does most of his crops, sometimes in Europe because organic markets are stronger there, and bread bakers still use millet and buckwheat.

Kirschenmann believes humans fell from grace ten thousand years ago, the fall a sin of pride that came from domesticating plants. Since then, all of agriculture has been an attempt to enforce distance from nature. By contrast, natural systems are diverse and feed themselves in cycles of growth and decay, and Kirschenmann makes a profit by reentering these cycles, because he doesn't buy fertilizer. He grows it; and he can sell high by selling food for which consumers will pay a premium price.

Buckwheat, beer, and a gong illuminate for me Kirschenmann's notion of harmony. We bounce along, jammed in the cab of his behemoth tractor on a bitterly cold morning. Water is standing in his fields, because of the strange, cold spring. He shows me a flooded plot that should be planted to wheat by now. His neighbors in the same situation will be down at the disaster office claiming crop loss and federal checks. Kirschenmann will get no federal check. He will simply wait for the weather to warm, at which point it will be too late to plant wheat but in plenty of time for warm-season crops like buckwheat.

As an organic farmer, Kirschenmann practices a subset of sustainable agriculture. Organic growers eschew all chemical fertilizers and pesticides. Sustainable ag can use chemicals, although it tries to limit them; but both schools believe in rotating crops, incorporating livestock into crop production, and integrating systems of managing pests, markets, and nutrient cycles. The antithesis is industrial ag. Nevertheless, in some of its advertising, ADM tries to nudge at least its image into the sustainable-ag column by backing a practice called "no-till," which does away with plowing and other forms of tillage. What no-till tends to mean in practice, however, is that weeds are controlled by herbicides instead of cultivation; so organic farmers mostly avoid no-till, not to mention ADM.

Not to mention subsidies. North Dakota's agricultural officials conducted a study that found that both sustainable and industrial ag can be profitable. The primary source of income for both was sale of crops, but the second most important source for sustainable ag was sale of livestock; for their more "efficient" counterparts, the second most important source of income was government payments. Narrowed to one or two crops, factory farming is rigid. There are no options for rolling with nature's inevitable punches. Federal subsidies cushion that rigidity.

Kirschenmann says the best thing that can happen to a farmer like him is the proliferation of brew pubs, a movement that has spread even to towns like Fargo. He can contract with the micro-brewery to raise brewing grains, sell them at a premium, and then sit with his neighbors on a Saturday night in a place where people make good beer. This is how markets build community instead of destroying it. We once came together over food, before it came from commodities.

Carolyn Raffensperger, Kirschenmann's wife, sounds a gong, then there is a moment of silence before we begin a dinner of their farm's beef. The quiet is complete on the prairie in Kirschenmann's earth-sheltered house. Out front an unpolluted slough full of ruddy ducks settles with the sunset.

ADM's corn is field corn (also called grain corn), not the kind that comes in cans or is eaten on the cob or popped. It is the hyper-hybrid descendant of the corn first domesticated in Mexico. For a long time its ancestor was eaten as food, and it still is in some remote villages, but not for much longer. As a result of NAFTA, the export of industrial ag to Mexico has greatly accelerated, a fact that has alarmed botanists. The typical Mexican corn farmer using traditional methods raises forty varieties of corn, and this gene pool is in danger. Like farming, corn itself has been narrowed until it no longer produces a food, a loss that is cultural as well. Those forty varieties each have separate uses, seasons, customs, recipes, and celebrations; the basis of culture.

Even the World Bank says NAFTA will drive as many as 300,000 Mexican farmers from simple but workable lives in the countryside to marginal lives in cities. In various guises, this same political-economic force is driving a worldwide urban crisis: Third World countries have seen a fivefold increase in their urban populations between 1950 and 1990. Virtually all of these refugees assume lives of the worst sort of poverty, and much of the blame can be laid on American grain. ADM stores and brokers much of the grain the United States dumps on Third World countries. Andreas was standing in the room when part of the law enabling this dumping was written. In 1978, ADM paid a $200,000 fine after pleading no contest to price-fixing in the Food for Peace Program. Between 1985 and 1995, ADM had received more than $130 million in export subsidies through the USDA.

U.S. grain, free or otherwise, puts Third World farmers out of business, sacking local agriculture and local markets. Case studies going back to the 1950s demonstrate this in India, Peru, Egypt, Somalia, Senegal, and Haiti. This is one way in which we pay to hide the surplus that we have paid our farmers to produce.

The most troubling surplus in all of this, however, is people—those who have moved off the land to the ring of house trailers around plains towns, or to sprawling slums in the cities of America and other countries. We justify our agricultural system by saying it feeds people, but how is it that we still can't answer the more fundamental question as formulated by the farmer and poet Wendell Berry? If the system can only work by making people redundant, then what are people for? he asks in an essay that bears that question as its title.

This question does not worry Hiram Drache, not even a little bit. Drache is an agricultural historian and an unabashed proponent of industrial agriculture. Vogel and Kirschenmann are subculture; Drache speaks for the mainstream. Literally—ADM hires him to lecture. To the outsider still holding a misty-eyed image of the American farmer, the cover of Drache's most recent book, *Legacy of the Land*, is antidote. The book cover shows an enormous, monochromatic and depopulated rural landscape, broken only by a metal-sided factory.

An engaging old man in standard northern-tier, Lutheran-issue blazer and tie, he holds forth from a warren of an office at Concordia College, in Moorehead, Minnesota, where he taught before retiring. He reminds me early in our discussion that there are now two counties in North Dakota whose county seats hold fewer than thirty-five people. He says this represents an "opportunity."

"I think we could be closing down a lot of county courthouses," he says. "With the computers we could send out the checks."

He believes computers will do most of the work of farmers, from satellite-based mapping of fields, to calculating and fine-tuning seed genetics and fertilizer doses (a capability already in use), to completely mechanized handling and packaging. The advent of this technology has been duly announced in a warm and fuzzy ADM ad that revolves around a conversation between a farmer-father and his farmer-son. Neither mentions that this technology is another nail in the coffin of the family farm.

Peter Bloome was assistant director of the Illinois Cooperative Extension Service when I interviewed him, and is now at a similar post in Oregon. An outgrowth of the Land Grant College system established by Lincoln, the extension services have been roundly castigated in recent years as shills for industrial ag. Bloome and others agree the charge fits, but the excesses of industrial ag have caused some defections, Bloome among them.

He says these high-tech systems generally are used by buyers such as ADM who contract with farmers to grow a specific crop, then use the technology to micromanage the farm. Control over decisions passes from farmer to contract buyer; the information moves to behind the wizard's curtain, leaving the farmer to serve as conduit and front man for lobbyists.

Industrial ag displaces people. Drache says North Dakota has a future dictated by the Wal-Mart philosophy—i.e., any town big enough to have a Wal-Mart will survive. The death of the rest he sees as no loss. Recalls Drache: "Once at a conference I heard a young woman say every small community that wants to should be allowed to continue. . . . I got up and said, 'I don't know when you have been in rural America last, but I have been through town after town after town where . . . I've said, Thank god I don't have to live there.'"

If you think him wrong on this count, you haven't been in rural America recently either.

The *Empire Builder* winds its way south from Fargo to the Twin Cities, both urban, polite, progressive, and clean. Yet in the Birkenstock quarter of St. Paul, in front of latte stands and bookstores, earnest young women in print dresses distribute leaflets denouncing hog factories. In Iowa, Minnesota, Illinois, Nebraska, Missouri, and especially North Carolina, the big sty state, public policy discussions have turned to hog factories. Said one Missourian of her state, "There are two hot issues: abortion and hogs."

We hide grain surpluses in foreign aid, but the more traditional way of hiding corn is in livestock. The first cattle drives just after the Civil War brought longhorns from Texas to the Midwest, where they were fattened up on a surplus of corn that has existed in virtually every year since. Thus began the habit that produces the fat meat at the heart of the nation's 700,000 annual premature deaths from coronary disease. It is nothing more than a habit; we can produce perfectly good (and better for you) grass-fed beef without corn, but worldwide two-thirds of the corn crop is fed to livestock, increasingly in the United States to swine. Grazers like cattle and bison could do a better job of producing our protein on natural grass systems, which, unlike plowed fields, can be sustainable. Hogs can't graze, but we and the landscape would be a lot better off if there were a lot fewer of them in our food chain.

The new wave in hog farming, what many believe to be the new wave in farming itself, began in North Carolina but logically spread to the corn/lysine belt. Corporations founded mega–hog factories holding twenty thousand to a hundred thousand animals. A single county in North Carolina holds more than a million hogs; they outnumber residents twenty-five to one. In Missouri, one county holds 1.5 million.

This trend is emblematic of the future of agriculture for three reasons. First, it is filthy. As noted, the average pig's gut produces about three times the waste of the average human. That county in Missouri with 1.5 million hogs warehoused in a few buildings finds itself with a sewage-treatment problem the equal of a city of five million people, but none of its waste is treated as sewage. Regulators now discuss hog farms in terms of "cleanup," a phrase once reserved for toxic waste sites, and they use this term with cause. In 1995, twenty-five million gallons of liquefied hog shit busted a lagoon and spilled into North Carolina's New River, wiping out seventeen miles of aquatic life. At least 88,000 fish died in a similar spill in Missouri, one of seven in the summer of 1995. Such spills are part of the record in Iowa and Minnesota, but those are the accidents. Even if all goes well, residents of communities that draw hog factories first get to live a few years with an unmentionable stench, then they get to leave.

Second, it is oligarchic. Small owner-operated hog farms can produce pork for about the same price as hog factories, pound for pound, but the larger operations can draw venture capital and negotiate better contracts with meatpackers. In the last twenty years, the number of hog farmers in the nation has declined by 68 percent, close to a half-million farmers put out of business by corporate farming. Now 4.8 percent of the producers turn out 43 percent of the nation's pork.

Third, the industry is a parasite, an outgrowth wholly dependent on corn made cheap by federal subsidy and on corn and soybean derivatives sold by ADM and its ilk. In nature, the creation of an enormous untapped biomass will surely lead to the evolution of creatures sufficiently large and adaptable to feed on it. The principle applies both to hog factories and to ADM.

The train rolls out of the Twin Cities, south to Chicago, and then a new train continues south from there into what was once tall grass prairie, now wholly surrendered to soybeans and the domestic tall grass, corn. The landscape here is different from high-plains wheat country. Here the towns are neater and closer together. You see picket fences. There's more money in corn. It is not so much food as it is a raw material (only about 7 percent of the crop is used in cereals and other corn foods like tortillas), and still it is our most lucrative crop. I leave the train and drive a few miles away from the main line, the last few miles across ground zero of American agriculture. This, too, is a filthy place.

A farmer here tells me his placid rural life means his wife will not drink well water when she is pregnant. In Decatur, the Health Department sometimes issues bottled drinking water for infants. Corn is grown with nitrogen fertilizer. In central Illinois and throughout the corn belt, many of the surface water supplies (including Lake

Decatur, which is in sight of ADM's headquarters) and some of the groundwater supplies have unacceptably high concentrations of nitrates from fertilizer runoff. Nitrates cause methemoglobinemia, or "blue baby syndrome," which can be (and has been) fatal. The floods of 1993 flushed so much nitrogen from the Mississippi Valley that it created an algal bloom in the Gulf of Mexico, a raft of gunk that turned out to be an ideal breeding ground for disease. As the raft floated up the East Coast, it spread a mutated virus that killed dolphins, beluga whales, Atlantic harbor seals, and porpoises as far north as the St. Lawrence Seaway. A similar event occurred in the nitrogen-fouled North Sea a few years before.

Virtually every stream in Illinois is contaminated with the four herbicides used to grow corn and soybeans, some above federal standards. An official of the Illinois Environmental Protection Agency says the situation is bad enough to warrant "enforcement action." Does this mean the state stops the corn farmers from putting pesticides in the water?

"Oh, no, that wouldn't be practical," he says.

It means the state makes communities find a new source of drinking water.

The corruption of life, however, is not simply local. Once we hid our corn surplus in livestock and impoverished foreigners, but our ability to produce corn has outrun the supply of these. What the situation required was a new idea that would treat all of us like livestock.

ADM makes most of its corn into syrup. Nationally, about 42 percent of the processed corn goes to sweeteners, something almost no one ever buys directly but which most everyone eats. Just read the label on most any processed food, especially "fruit juices." Look for "high fructose corn syrup" as the second ingredient, right behind water. It's in scalloped potatoes, barbecue sauce, salad dressing, ketchup, oatmeal cookies, Wheat Thins, Campbell's Chunky Soup, granola bars, canned fruit, SpaghettiOs, ice cream, and virtually every carbonated soft drink.

This ubiquitous "food" has spawned a wave of obesity in the

United States, a trend directly tied to decisions made in Decatur. To quote *Barron's*:

> In the late Seventies and early Eighties, ADM gambled that fructose corn syrup would eventually supplant sugar as the sweetener of choice for the soft-drink bottling industry. To overcome bottler resistance however, ADM had to constantly add new capacity in already glutted syrup markets. . . . The displacement of sucrose took place in agonizing stages. But ADM emerged the big winner, leaving rivals such as CPC International and American Maize in the dust.

As the freeway nears Decatur from the east, an audible "Jesus" resonates through the rental car. Then I realize it came from me and was provoked by my first glimpse of the panorama to the south—a horizon of steam vents, smokestacks, pipe, and steel. This is the Emerald City, the center of ADM. Here's where it's done, the processing, both seen and unseen, of politicians as well as of commodities.

Sugar, pure cane sugar, costs about twelve cents a pound on the world market, but twice that in the United States because of protective quotas set in a 1981 sugar bill heavily lobbied for in an effort funded by ADM. The cost of corn syrup hovers about halfway between world sugar and protected domestic sugar, a price differential designed to "overcome bottler resistance, a reluctance, it turns out, solely based on price." In 1996, when Congress supposedly revolutionized farm policy in a new farm bill, sugar was again controversial, but the provisions affecting that commodity survived virtually unchanged from previous legislation. The beauty of this, from ADM's point of view, is that the company's efforts leave no fingerprints. No money goes directly to ADM; it goes to sugar farmers. A congressional report summarizing lobbying on the 1996 bill doesn't even mention ADM's role (nor did an exposé of the sugar lobby on *The NewsHour with Jim Lehrer*); but, then, as *The New York Times* pointed out, "Archer-Daniels does not have a lobbyist in Washington;

it really does not need one." Still it gets one simple law that is the necessary and sufficient cause of making saleable corn syrup, the product that alone accounts for more than a third of ADM's profits, the single biggest contributor to its net gain. Meanwhile, this quota costs a family of four about sixty dollars a year in higher prices.

Corn syrup's price, however, may not be an altogether organic artifact of the market that quotas have artificially sweetened. At least that's the suspicion that drove an FBI investigation in the 1990s into allegations made by Mark Whitacre, a former ADM executive turned whistle-blower. ADM, along with competitors, controls about 83 percent of the corn-sweetener market, and Whitacre alleged that ADM conspired with foreign companies to set the price. Whitacre, however, lost his immunity from prosecution and his credibility when he was found to have bilked ADM out of $9 million. ADM did plead guilty to two charges of price-fixing related to those allegations and paid $100 million in penalties. Both Whitacre and Michael Andreas, Dwayne's heir apparent, did prison time.

Some of the smokestacks on the horizon mark the production of ethanol, of which ADM is the nation's largest purveyor. Ethanol is an interesting substance—an alcohol distilled from corn, but as much a distillation of politics. Promoted as an alternative motor fuel for a century, it is enjoying a recent boost in its political octane.

Because it derives from corn, ethanol production requires the energy and petrochemicals that feed tractors, transport trucks, and distilleries directly, and fertilizers indirectly. In fact, several studies that have examined the issue conclude that the production of ethanol consumes more energy than it yields. The more optimistic appraisals find a net gain only when the value of co-products such as corn oil are taken into account, and even then a unit of energy is consumed for every 1.24 units of ethanol energy produced. The most favorable says it takes about 1.3 units of energy to produce one unit of ethanol energy. This real-world mathematical deficit must be papered over with subsidy. Every dollar of profit ADM makes on ethanol costs American taxpayers eleven dollars.

If there were a contents-labeling law for politicians, corn money would be listed as the first ingredient. When Bob Dole represented Kansas in the U.S. Senate, he sponsored fifty bills supporting ethanol. *The New York Times* called Dole "ADM's staunchest ally on Capitol Hill." Noted the *Wall Street Journal*: "In the Senate, Mr. Dole has been the chief promoter of the ethanol subsidy."

Dole is the most notable beneficiary of the four million dollars ADM, Andreas, and family members have contributed to various political candidates over the past couple of decades. He has flown on ADM aircraft at least twenty-nine times since 1993. ADM sponsored Dole's commentaries broadcast over the Mutual Radio Network. Dole bought an apartment from Andreas in Florida in 1982 at less than market value. After Elizabeth Dole became the head of the Red Cross, Andreas's nonprofit foundation gave one million dollars to it.

About the same time that then-president George Bush proposed standards favoring ethanol, he received a $400,000 check from ADM. In 1994, Bill Clinton got a $100,000 check. Andreas himself organized a fund-raiser that netted the 1992 Clinton campaign $3.5 million, but at least this seems more refined than Andreas's earlier political adventures. In 1972, he walked into the Nixon White House bearing an unmarked envelope containing a hundred thousand-dollar bills. He ensured his place in history by writing the check that tied the Watergate burglars to the Nixon campaign. In 1992, Andreas-related checks were the top money source for the Republicans, and the third-highest for the Democrats.

Days after receiving the $100,000, Clinton ordered that ethanol be added to 30 percent of the gasoline in the nation's nine most polluted cities, despite information from within his own administration that showed gasohol causes new pollution problems. This proved to be a repeatable pattern. ADM was a leading donor to George W. Bush's campaign in 2000, a favor later repaid when one of the first acts of his administration was to refuse to grant California a waiver to rules requiring oxygenated gasoline. California argued it could meet federal clean-air standards without additives to gasoline, of which there are two. The state had already banned one of those, methyl tertiary butyl ether (MTBE), on the grounds that it posed health hazards, leaving only one alternative: ethanol. The ban on MTBE, however, became suspect when *The Wall Street Journal* reported that California governor Gray Davis met with ADM officials and took a $135,000 campaign contribution from Martin Andreas, a member of ADM's controlling family, before announcing the ban.

A longtime Decatur resident tells me that as a kid he loved corn bread, but he can no longer eat it, or anything else with corn or corn syrup in it. Like many hereabout, he has developed allergies as a result of the high concentrations of corn pollen in the air. The corn-pollen index appears on the local news during the growing season. He also tells me there are no hotels in downtown Decatur; like much of the Midwest, it has surrendered to chain-strip sprawl. Besides, the town stinks of cooking corn mash. And he tells me that ADM, a multinational with $12.6 billion in annual sales, has a public relations department of one person, and if he won't talk, there will be no talking.

I check into a chain motel and call ADM's lone-ranger PR department to ask for a tour. (I figure cornering Andreas is out of the question and have no desire to participate in a corn-belt remake of *Roger and Me*.) A secretary answers. No, he's out. I identify myself and lodge my request.

"Perhaps Martin Andreas will speak with you," she says. She'll check and get back with me. Sure. Martin is a senior vice-president in a flowchart liberally seeded with the Andreases.

I leave my room for half an hour and return to a message. "Andreas called," it says. I call the main switchboard and say only that I am returning Martin Andreas's call.

"I'll get you his office."

A secretary answers. I identify myself. She says he will speak with me right now. Then a voice snarls, "Andreas."

"Martin Andreas?"

"No. Dwayne Andreas."

And, yes, Dwayne, recently beaten in the press in the wake of the price-fixing scandals, and just as recently denying interviews, says he will grant one. What is an interview going to be like with the legendary CEO, bagman for the Nixon White House, and seventy-eight-year-old autocrat with a sufficiently hands-on approach to the corporation that he is only a call removed from the main switchboard?

"I'll tell you, he's a hell of a guy," says Emmett Sefton. "I guarantee you he'll have a nice suit on and you won't put anything over on him."

Sefton himself is a hell of a guy, a straightforward industrial-ag, corn-and-soybeans farmer who spent five years on the American Soybean Board. I speak with him in the neat, modest farmhouse on his thousand-acre farm a few miles from Decatur. He's reasoned and thoughtful about industrial ag, pensive even when he talks about farming being a lot less work and a lot more stress than it was a generation ago when he grew up on farms that still practiced crop rotation and had livestock. And, yes, it's common to hear that farmers don't want their kids going into farming, but one of his sons has decided to anyway. He talks about what he has done to better manage chemicals. (He once accidentally dumped a gallon of pesticide into his face.)

Sefton sees Andreas as the point man of industrial ag, the champion of farmers, and he's glad the patron wears a nice suit. He is the conduit to the international markets and the grease in the political system that makes the mountain of corn and soybeans disappear. Andreas has taught his protectorate some lessons about what he calls "tithing."

"I donate to both parties," says the commodity farmer who estimates that 15 percent of his income, as much as 40 percent in some years, comes from direct federal payments. "Any politician needs help, whether he requests it or not."

"If we didn't do something to move our product, we'd be in a hell of a shape today," he says. "I'm tickled to death ADM is where they're at. I can see their smoke every morning."

The stacks represent a simple system for Sefton and all corn growers. He raises a crop, often under direct contract to ADM. He loads it on a truck or train and ships it straight to an ADM elevator. A check comes. If the price is not large enough to ensure his profit, a second check comes from the government to make up the difference, guaranteeing that his corn will be there—and be there cheap—in years to come.

ADM has messengered me a five-pound press kit, and in it is an Andreas quote, something he once said in advance of an appearance before a group of Boston financial analysts: "Getting information from me is like frisking a seal."

There's a bronze statue of Ronald Reagan in the parking lot in front of the blocky, utilitarian office building centered in the sprawl of steam and stacks, but ADM's iconography is at least ecumenical. The 1995 annual report features a picture of John F. Kennedy on the cover, complete with the "Ask not . . ." quote. (The same document reports a net profit of $795 million; ADM has not suffered unduly in our nation's service.) Dwight Eisenhower gets prominent play inside, and, just to cover all bases, a particularly pithy quote is set in the shape of a cross.

On the top floor is the hushed, wood-paneled, memento-filled office of Dwayne Andreas. From what I know of suits, his is okay. His answers are rote, and he seems to pull them from someplace off in the middle distance. He's a little guy, and he seems tired.

"The free market is a myth. Everybody knows that. Just very few people say it. If you're in the position like I am and do business all over the world, and if I'm not smart enough to know there's no free market, I ought to be fired.

"The reason we don't call it socialism is that socialism is a bad word."

A bad word especially to Republican politicians who fuel their crusades for the free market on Andreas cash. Andreas tells me this is not about ideology.

"I don't talk to politicians about my business. Ever. I systematically avoid that," he says.

Then you just give them money? No strings?

"I never give money to politicians."

You don't?

"No. I respond. We have people all over the company that respond to fund-raisers. We have businesses in three hundred locations, so naturally we get drawn into supporting various political candidates. But I do not deal with political candidates at all."

You don't?

"No. I never discuss my business with them and never make any contributions to them. . . . There is no connection between campaign contributions and political activity. . . . Who doesn't respond to their favorite politician? If you're any kind of a citizen you respond. Otherwise how the hell would you have a democracy? The poor devils can't run for an election unless they can at least buy a poster. Let alone television."

But Sefton says you're important in politics because you are the farmers' advocate.

"I don't have to be an advocate of anything. If you're in the business you have to be an observer. I observe what is happening. I don't have to advocate one way or the other. And besides, what difference would it make if I advocated or didn't advocate?"

Some of the politicians Andreas doesn't talk to and doesn't give money to have passed a new farm bill with the stated intention of ending federal farm subsidies in seven years. They were emboldened in this endeavor by the fact that commodity prices are at the apogee of their orbit, and farmers don't need the subsidy just now. Does Dwayne Andreas think for a second that farmers will remain unsubsidized when prices fall, as they surely will?

"No."

Which is precisely the unequivocal answer that Sefton and every other farmer and farm expert has given me to that question during two months of interviews.

"You can't have farming on a total laissez-faire system because the sellers are too weak and the buyers are too strong," says the world's strongest buyer.

What about the co-ops springing up in North Dakota and elsewhere? Might they make the sellers strong enough to stand up to ADM?

"Co-ops own more than twenty percent of our company. Overseas co-ops own half. A million or more farmers are either partners or shareholders in ADM. We're really an extension of the farm. It's all one great big machine," he says.

North of Decatur, up on the freeway exit strip, I pull into Applebee's, or some other cookie-cutter "neighborhood" restaurant. Does it really matter which one? This is the corporate clutter novelist Tom McGuane calls "the escalating boredom of the monoculture." The waitress extrudes from her Gap khakis in a way that detractors of midwesterners once called "corn-fed," only now it is a better joke. We all are corn-fed, because we all eat identical products off identical trucks at identical strip malls, coast to coast. I order my slice of the collective pie and finally get the ADM slogan, "Supermarket to the World." Not supermarket as in Safeway. After all, ADM doesn't sell anything directly to consumers. It doesn't even advertise in the traditional hardball, product-pushing sense (although there are all those sponsorships, euphemistically called "underwriting," of political talk shows and public-broadcasting news programs).

It's "super" as in *über*, as in "above all." Above all markets and above all people.

The interview with Andreas has been a failure, although by the two current standards of journalism, not really. One standard approach is simply to ignore ADM and the pernicious effects of corporations on American agriculture, an effect particularly pronounced at National Public Radio and on *The NewsHour with Jim Lehrer*, both of which enjoy ADM "underwriting" and run corporate-ID statements that look and sound like straight-ahead commercials. (Control over media, however, can be even more direct. In 1994, Andreas bought $9.5 million worth of American Publishing Co., owner of the *Chicago Sun-Times* and 340 other newspapers.)

The second, more prevalent, approach is the "gotcha." This entails compiling a laundry list of misdeeds (like Andreas's role as Watergate bagman, ADM's antitrust violations, the FBI investigations, and the like), confronting the man himself, and then dancing around the denials. Like the food in this place, it is shit—processed entertainment at best, that gives the illusion of reporters having done their jobs by policing the rare cases of excessive behavior. The miscreants (defined as those who speak their intentions outside of their private clubs) have been dealt with, and all is well.

The thing is, I want it all. I want to unroll that social landscape from the Rockies to the Ohio River valley, the dream of hard work and honesty it was, and the oligarchy it is now. I want to replay every blank fat face staring at every blank plate in this place and say, "Mr. Andreas, how did we get like this?" The thing is, I think he knows.

A longtime observer of agriculture told me, "You'd have to stand in line to hate ADM," but standing in that line would be a mistake. ADM, thanks especially to Andreas, is a lightning rod. He is nineteenth-century enough to advertise his empire-building. He buys his politicians at public auction—all the better for Cargill, ConAgra, General Food, Borden, Continental Grain, CPC, Ajinomoto, American Maize, A. E. Staley, and the rest who buy silently.

Gayle Goold, a corn farmer, a sustainable ag man, tells me there's a big problem in Illinois, that a corn worm has learned the evolutionary trick of laying its eggs in soybeans, thereby defeating crop rotation. That's a parasite's job, to look for a big, untapped pile of biomass and then evolve to tap it. Andreas has told me a true thing. His job is to observe, to respond, to evolve, to co-opt the co-ops, to morph, to feed on a decaying system.

I ask Goold what the rest of us, the 98.4 percent, can do about this, and he says, "Grow a tomato."

Our people have no idea what real food tastes like. The effect of real tomatoes on the American public could very well work like (and now I can only imagine) a taste of integrity from our politics. Quality is subversive. The farmer is saying: We are what you eat.

WHY WE ARE WHAT WE EAT

It is fall in Montana, with low, slanting, afternoon sunlight and evenings tinged with the nip of frost. The languorous days of summer are surrendering to the coming season, another year folding into the longer cycle of my own life.

There are ways to celebrate such days. If I am lucky enough to have shot a few pheasants, I smoke them on a grill all through a long afternoon, and hope that the frost has not yet stopped the squash and tomatoes ripening on my garden's vines, that a few ears of fresh sweet corn remain. There will be music, hours spent in the afternoon sun with my 1926 Martin guitar and all the fully aged tone I can pull from it, a profound pleasure, and all the stories I can't pull from it, a tantalizing vexation. I think of a nine-thousand-year-old flute archaeologists recently unearthed in China. It was made from the leg bone of a crane, and it could still play the A-major scale it was built to play, in recollection of the crane's call. The maker probably hunted the crane as I hunted the pheasants, so I play an A-major scale.

Toward sunset, the food is ready. My wife and I sit, pour wine, and watch the last bit of light shine in our glasses. My attentiveness to the wine's rich, voluptuous red is far more anciently encoded than my ear for an A-major scale. There is no telling where the food and music end and the sex begins, but it has always been so. Human taste buds, lips, tongues, noses, and genitals all are wired with the body's most hypercharged sensory equipment, the sharpest cells, called Krause's end-bulbs. Night deepens on our Montana hillside and we bask in that rarest of human luxuries, pure silence, pure dark, sensuality.

I insist on sensuality. I guard my smoked pheasants, old guitars, and quiet as jealously as any miser guards gold. They can do far more to protect me from what we humans have become: insensate, insensitive, inhuman. For the millions of years of evolution that made us, the ability to fully sense food and sex was the foundation of our humanity and the core determinant of survival. For ten thousand years, those same pleasures have been reserved for a few of us. Complete indulgence of sensuality is rare, and, as a rule, the purview of the rich. For ten thousand years, *Homo sapiens* has been unable to take its humanity for granted. Those who would resist dehumanization do so by daily staking a claim to it, by self-consciously adopting an aestheticism our hunter-gatherer forebears practiced by simply living. With the advent of agriculture, those qualities that united us—in fact, quality itself—came to divide us. Civilization did indeed modify the human genome, but only slightly, only around the edges. We remain at our genetic core largely what our hunter-gatherer history made us, which is to say, sensual beings. All of humanity at some level still requires the aesthetic. What was invented with civilization was the ability of some to deny sensuality to others.

Because Western tradition pays most of its attention to Western tradition, our foremost examples of the ancient elitism of cuisine comes from the Romans. Details of their banquets are well recorded and, in line with the rules of Krause's end-bulbs, read like descriptions of orgies, because they were. In her hypnotic *A Natural History of the Senses*, the poet Diane Ackerman writes:

> Romans adored the voluptuous feel of food; the sting of pepper, the pleasure-pain of sweet-and-sour dishes, the smoldery sexiness of curries, the piquancy of delicate rare animals, whose exotic lives they could contemplate as they devoured them, sauces that reminded them of the smells and tastes of lovemaking. It was a time of fabulous, fattening wealth, and dangerous, killing poverty.

The poor were but servants at the banquets.

This arrangement was by no means original to Rome; it already had been well established in the centers of agriculture. By 500 B.C., the kings of the Persian empire were commonly dining daily with about fifteen hundred guests, the nobility of that society. They ate a wide variety of animals, including camels and ostriches. They employed specialized chefs, including bakers, pastry makers, drink mixers, and wine attendants. They drank from gold and silver vessels. One king, Darius III, had as many as four tons of such vessels. Meantime, the diet of the country folks at about the same time and place was "simple and monotonous," according to one food historian, consisting almost solely of dates and milk from camels, goats, and sheep.

Similarly, the nearby Turkish monarchs thought so highly of cuisine that a pot and spoon became the symbol of their elite military forces, an emblem of a higher standard of living. The Ottomans maintained royal households with four sub-kitchens: one for the king alone, a lesser kitchen for nobility, one lesser still for the harem, and the simplest for the palace staff. At times the Ottomans officially stratified breadmaking, with the most refined flour reserved for the sultan's bread, middle-quality loaves going to officers, and the coarsest dark bread left for servants.

This hierarchy of cuisine remained the rule in Europe, typified by an Ottoman-like stratification of bread. As the food historian Kenneth Albala writes, "Because it has been the staple of the West, bread preferences are almost always an encapsulation of social climate. In fact, at times, the whiteness and texture of bread have been arranged hierarchically and have matched precisely the structure of society." But for the peasantry, bread—or one of a few other carbohydrates—was all. The writer Larry Zuckerman observes in *The Potato* that for French peasants, even into the twentieth century, "Grain was their sustenance, and when it failed, they starved."

Peasants, of course, also grew the produce for the tables of the rich, and so had firsthand knowledge of what they were missing. Zuckerman cites the memories of French novelist Pierre Gascar, whose grandmother produced goose livers for the pâté of upper-class tables:

Following the time-honored practice, she force-fed the birds so that their livers would enlarge. Cruel as this was to the animals, it was also a painful irony for herself, because she performed the chore throughout a life during which she seldom satisfied her own hunger. "It makes you pitiless," [Gascar] wrote, "when you have to force-feed an animal: It is just, in your eyes, that satiety becomes a punishment."

Sumptuary rigidity held sway among native agriculturalists as well, arguably more so. The Aztecs were the gourmands of the New World; besides gold, Spanish conquest gave the rest of the world maize, tomatoes, squash, chilies, beans, turkey, a long list of fruit, and, above all, chocolate. Yet the conquerors' reports of the diet of common folk among the Aztecs speak mostly of maize and a few local meats on feast days, while the accounts of feasts in the court of the emperor Motecuhzoma II are wide-eyed and detailed. The Franciscan friar Bernardino de Sahagún was so taken with what he saw that he took great pains to catalogue Aztec foods, devoting pages to a single banquet. As a result, we have a fairly complete accounting of the royal diet. Maize was a part of it, but only the whitest of maizes. There was a wide variety of fruits, elaborate sauces and spices, and meats, including waterfowl, turkey, venison, and human flesh. The friar dedicates pages to descriptions of casseroles alone, detailing ingredients such as fowl, red chilies, tomatoes, ground squash seeds, yellow chilies, fish, frogs, tadpoles, locusts, maguey worms, shrimp, and unripe plums. He found tamales made of frogs, tadpoles, mushrooms, rabbit, pocket gophers ("tasty, very tasty, very well made, savory, of very pleasing odor"), fruit, beans, and turkey egg, as well as salted tamales, pointed tamales, white tamales, white fruit tamales, red fruit tamales, tamales of tender maize, tamales of green maize, brick-shaped tamales, plain tamales, honey tamales, bee tamales, squash tamales, and maize flower tamales, to name a few. All of this flowed from a complex trade network in conquered regions. Tribute came in the form of a variety of goods, but especially elaborate foods to feed the court.

Friar Bernardino claimed that chocolate, which the Aztecs mostly took as a drink, was reserved for royalty, specifically male royalty. (The women ate separately, and at the point in the banquet where men drank chocolate, the women's table got a bit of grain gruel.) Other writers suggest that cacao was so widespread across the Aztec tribute network, which ran south at least as far as what is now Guatemala, that common people must have skimmed a bit along the way. Regardless, chocolate remained a marker of status. Individual cacao beans, in fact, were used as money.

The Spaniards were quick to adopt this strange new drink, and it almost immediately became fashionable among the nobility of Europe. The chocolate fad became so widespread that bishops had to issue decrees forbidding people from drinking chocolate during Mass.

There are some exceptions to the practice of culinary elitism in the civilized world, instructive for what they say about the symbolic importance of food. It was an Arab tradition, for instance, for servant and master to dine together, an expression of a fundamental egalitarianism among early Muslims. Japanese rulers in the earliest part of that island's civilization eschewed elaborate food, adhering to long-established Buddhist asceticism that, by limiting the attention expended on food, implicitly acknowledged its power. Japan finally did develop an elaborate cuisine during the Edo period, which began in 1639, and most modern Japanese dishes date to then. It developed not among the aristocracy, though, but among the merchant class that arose during that period, a new cuisine tied to a new elite.

The most revealing of the exceptions, however, arises in the Christian tradition and also is rooted in asceticism. The early Christian Church in the Middle East was a slave-class movement, a counterculture to Roman opulence. Thus, as the Romans indulged their bodies, the slaves made a virtue of simplicity and denial of the pleasures of the flesh, developing the body-soul dualism unique to Christianity. In the extreme—and there were many periods of extremism throughout Christian history—it crescendoed into a loathing of the body, expressed in hair shirts and self-flagellation. The drive toward asceticism did not thwart the development of cuisine, but kept pace with it. By the third century after Christ, the church fathers were officially endorsing fasting as a means of atonement, as well as what would become the abstemious monastic diet. The practice of forgoing meat on Fridays has only recently faded, and Lent is still with us. So is Mardi Gras, though, and so is the tension inherent in all of this indulgence and abstention. The body remains, and won't be denied.

The feast is probably older than the fast, but just as the fast arose as a religious counterweight, so in turn did features like gluttony, drunkenness, and sexual license take on feastly importance. The word "carnival" is rooted in the Latin word for meat. Kenneth Albala writes:

> The most universal feast was held on the day before Ash Wednesday, Martedi Grasso, or Mardi Gras, when all meat and eggs had to be consumed before Lent. This day of meat eating or "Carnevale" often became the occasion for gross indulgence. Drunkenness, flesh eating, violence and sexual license were all associated with this binge preceding the rigors of abstinence. By the late Middle Ages, mock battles would be held between personifications of Carnival and Lent, and the natural order of society would be subverted in mock trials, mock weddings and even mock prayers. Indeed, the world was said to be turned upside down. Gluttony was still considered among the seven deadly sins, though this rule, too, was momentarily suspended.

The best example of Christian dualism at work may well be the Benedictines, an order of monks who at times backslid so thoroughly from the monastic diet that outsiders complained of their inappropriate corpulence. There was some basis for the complaints. The Bendictines spread viticulture through Europe, developing along the way renowned cheeses, pastries and confections, cordials, Chartreuse, vermouth, and champagne (Dom Pérignon). In reaction to such excesses, new monastic orders steeped in asceticism arose, perpetuating the dichotomy within the Church. The asceticism survived in several strains of Protestantism, including the Shakers, the Quakers, and the Puritans, a group that banned all spices on the grounds that they were sexually arousing. Writes Ackerman, "Food has always been associated with cycles of sexuality, moral abandon, moral restraint, and a return to sexuality once again."

We use food to set ourselves apart, on religious and moral grounds, certainly along economic lines, but also ethnically, as is nowhere more clear than in the great melting pot, which has not melted as much as we think. Hispanics were called "beaners" just a generation ago, the French were "frogs," Germans "krauts," and Italians "spaghetti snappers" or "macaroni heads." Among bikers, an Asian-made motorcycle is a "ricer." At the same time, the 1950s era American obsession with the notion of "sanitation" and the hyperprocessing of food beyond anything recognizable as once growing or living was a way of removing food—and oneself—from barbarism.

There is nothing new in this. Betty Fussell, in her book *The Story of Corn*, notes that nineteenth-century settlers in the American Midwest considered the wheat they brought with them from Europe clean and fit to be made straightway into bread, but settlers, unlike the natives, processed indigenous corn into something unrecognizable (as Americans still do). Writes Fussell:

> My grandfather prayed in words, because dancing, in his language, was a sin. So was the body, so was all matter. Flesh and blood were of the earth, earthy, corruptible and corrupted, not God's but the Devil's work, awaiting the lightning blasts of the Apocalypse and the trumpets of Armageddon. The split between body and soul was as complete as between man and God, Indian and white. In their westward migration, my tribal Presbyterians were aliens among the heathen and fundamentalists among the heretic. As self-willed exiles, they could not comprehend a people whose gods were rooted in a particular earth, both patch and globe, as strongly as stalks of corn.

Nor were such sumptuary distinctions between civilized "us" and barbarian "them" limited to the West. In fact, long before the Persians were gourmands, the Chinese linked elaborate cuisine to the heart of their civilization. Chinese mythology held that preagricultural people ate raw flesh, like barbarians, and so cooking became the very definition of civilization. The more elaborate the preparation, the greater the distance from barbarism. In the sixteenth century B.C., the emperor Tang appointed a cook as his prime minister, and thereafter the proper seasoning and handling of food became a metaphor for good government. One Chinese expression for "seasoning the soup" came to mean "to be a minister of state." And of course the most sophisticated dishes were concocted for the most elaborately civilized of Chinese, the emperor.

The us/them distinction has played out regionally as well, reflecting and reinforcing regional, political, and ethnic tensions. China, for example, has throughout its history been more or less divided into northern and southern regions that coincide with the country's two main agro-ecological areas, one growing primarily wheat, the other rice. As the capital periodically shifted between north and south, Beijing and Nanjing, the change in regimes run by northern or southern people brought with it a shift in the ability to set fashions and tastes for the elite, especially in cuisine. Food historian Françoise Sabban writes:

> The struggle for preeminence has always been mirrored in the cuisine and food habits of the Chinese. A political exile, for instance, might evoke a specific food to signify his homesickness, and indignation about the injustice he is suffering. Or again, one who has betrayed an unfamiliarity with foods peculiar to the "other" China gives a reason for others to stigmatize his or her ignorance, or naiveté or haughtiness. The strong allusive value of some foodstuffs that assumed the rank of regional emblems thus made deep or nuanced comparisons unnecessary.

This use of cuisine to divide arises most clearly in the form of food taboos, which are not wholly cultural artifacts but grew out of the omnivore's dilemma: we needed to range afield to try new foods, but some of them were toxic. So we learned to identify the toxic ones. This helps to explain both our attraction to sweet foods, sweetness being a signal for carbohydrates and dense energy, and our aversion to bitter foods, bitterness being a signal for toxins. At the same time, we revel in some bitter foods, such as coffee, tea, and alcohol. Why? One answer I imagine is that if it is the omnivores' dilemma to live on the edge, we have internalized a certain pleasure in beating the odds, in winning at dietary Russian roulette. Consuming certain bitter foods is a way of celebrating a victory over our environment. Biology offers corroboration for this theory: some bitter foods provoke a powerful endorphin reaction that heightens our sense of taste. Even as our defense systems shout, "Be careful, this is bitter," the endorphins make us more alert and attentive, and increase our sensual enjoyment as a side effect.

Yet our information about toxins is not entirely the product of personal experience. Often people discovered a given food was toxic because someone else died from it, and others were told. In other words, we are deeply conditioned to pay attention to cultural information about food. If an authority figure tells us a food is unclean, we believe it. This authority can be used for practical purposes. For instance, the raising of swine for food was banned in the Middle East, as, more recently, was coffee use among Mormons, because omnivorous swine ate scarce protein, and coffee, which had to be imported in nineteenth-century Utah, took a heavy toll on resources. That is, the taboos were used to enforce a social good not rooted in avoiding toxins. Taboos could be useful, but just as often they were used to divide people.

Taboos preceded agriculture. Among surviving hunter-gatherers, there remains a widespread belief that we become like what we eat. By consuming a certain animal, the theory holds, we too take on qualities of that animal, which explains why there is still a vigorous Chinese trade in tiger penises. Just as easily, we can become disgusting by eating something disgusting, a useful way to render a whole gender or race or ethnic group or caste disgusting in the eyes of those who do not share a certain habit of diet.

There may be a rationale behind many taboos, but it is hard to imagine what it might be in the case of, for instance, the Hindu prohibitions against all foods offered by an actor, artisan, basket maker, blacksmith, carpenter, cobbler, dyer, eunuch, goldsmith, harlot, hermaphrodite, hunter, hypocrite, informer, jailer, leaser of land, manager of a lodging, menstruating woman, miser, musician, paramour of a married woman, person who is ill, physician, police officer, ruler of a town, seller of intoxicating beverages, spy, tailor, thief, trainer of hunting dogs, usurer, weapons dealer, or woman with no male relative, as enumerated in the Dharma-Sutra.

Food divides. In its mildest form, it simply reinforces ethnic identification, but the impulse to divide seldom limits itself to the mildest form. Hindus reinforce a brutal caste system with food taboos. Tribalism, racism, prejudice, even genocide are coded in food, and these, unlike poverty or disease, are not artifacts of agriculture. Recall the bands of male chimps wiping out all of their counterparts in a neighboring band. Throughout human history, even beyond human history, we primates have been vicious to any of our own species we could define as "other." The line dividing us and them can be as arbitrary as skin color, religion, political allegiance, or, in modern times, loyalty to a particular football team. This racism is the downside of what is nonetheless a useful trait in our existence. Evolution tells us it must be useful, or it would no longer be with us. These lines of prejudice are how we reinforce the internal ties of community that hold us together; they are the flip side of social cohesion. We are able to define and defend ourselves as a group by using the concept of "other" to draw the line around us. Food plays a defining role in the cohesion as well.

In contemporary life, the assertion of dietary distinction can be a way of marking one's individuality within the group, or one's allegiance with a subgroup. Is there an American urbanite alive who has not waited to order coffee while a twenty-something at the head of the line loudly asked, "Can I have that with soy milk? Was it made with organic bananas? Was the coffee shade-grown? Is the carrot cake vegan?" This is no different from young people's adoption of other on-the-edge affectations—from piercings and shaved heads to loud music and black clothing—that paradoxically reinforce their place within a group even as they proclaim their rebellion. How many mothers have heard a teenage daughter (vegetarianism is more common among women) announce that she will no longer be eating the "poison" (meat) served at her family's table, thereby announcing her emergence into full-blown adolescence? And a mild form of distinguishing herself it is, considering the self-defining and self-defying food maladies available to the modern American. (This avenue is not restricted by age or gender. Once, during lunch break at a weeklong workshop I was attending along with fifty other adults, a man in his fifties, a skinny guy with a beard and long straight hair, marched to the table, examined the goods, and announced to the entire room, "There is vegetarian lasagna here for anyone with a conscience.")

In the worst cases, this sort of line-drawing reinforces racism and discrimination, while, in lesser cases, it stands as one of modern society's irritants. This power of food to define is not, however, all negative. We all need to define ourselves, to have a sense of identity.

The Mayan people knew where they came from: they were created by maize, and not just in the sense of "you are what you eat." The Mayan sacred text, the Popul Vuh, said unequivocally that the gods manufactured the first humans from maize dough.

Likewise, the Pueblo Indians to the north knew of their own maize lineage. They descended from the seed people, particularly the flesh of seven sisters: yellow, blue, red, white, speckled, black, and the youngest, sweet corn. Flute music summoned these seven to gather for a night's dance; then they departed, leaving their flesh behind. Maidens selected for Aztec sacrifices spent the month before their final day performing a dance in which they shook their hair across their breasts like corn silk, a dance that urged the corn to grow.

Confucius wrote that all he required in life to make him happy was a little rice and water. Christians' fundamental plea to their god was "Give us this day our daily bread," and it was not metaphor. They meant bread. Distinctions between royalty and commoners aside, mainstream agricultural people came to define themselves, their very creation and essence, as identical to their central carbohydrate: wheat in Eurasia, rice in tropical Asia, maize in America. This translated to true reverence for the food in question. In places in Eastern Europe, there were still people in the twentieth century who practiced the long-standing custom of kissing a piece of bread that has been carelessly dropped on the floor. Scrap bread was never thrown to the garbage but fed to animals. Aztec women breathed gently on maize headed for the pot so that it would not fear the fire. This reverence is deep and fundamental, just as it is defining of a person's culture. Food carries culture.

In the Sierra Madres of northern Mexico one finds entire villages where Nuatal, an Aztec dialect, is still the principal language. Once when I was visiting such a village, a farmer—a poor man, really—invited me into his house. Its roof was supported by a central post that had become a sort of shrine. On it hung a drawing of his household's saint and a big bundle of ears of dried corn, his seed corn saved from that year's harvest. The line of corn was as entwined with his family's line as was the household saint.

The Mexicans in these villages are indeed poor in one sense, but not in another. As noted, a farmer here grows as many as forty different varieties of corn, largely still sorted along the same lines as the Pueblo's seven sisters: yellow, red, black, blue, white, speckled, and sweet. Each has its season. Each has its preferred place and time of growing. Each is tied to a particular recipe, so a single species sponsors a varied cuisine.

This, however, is only the beginning. These same farmers grow several varieties of squash and beans, and a long list of greens derived from wild chenopods. On a single day, a single farmer took to market plums, black cherries, pears, chilies, prickly pear cactus fruit, amaranth, squash, radishes, lemongrass, thyme, avocado, mamey, fava beans, sweet potatoes, and blackberries, all from a plot of perhaps two acres. Presumably her own table is as varied. The land around grows coffee, turkeys, vanilla, pimienta, plums, nuts, quince, strawberries, and marijuana. The people here use something like 250 species of edible plants. Poor as they may be, these people hate the thought of leaving their villages and going to cities, where they will have to eat what city people eat. City dwellers do not have access to the variety of foods that this rural life has maintained.

In a sense, this is what biologists would call a refugium, a place where a suite of plant and animal life can survive in the face of natural upheaval like glaciation. This place is a refugium against the dulling monotony of catastrophic agriculture. If agriculture is culture, this is subculture in a truer sense than is signified by body piercings, black clothing, and cacophonous music. It maintains a full complement of foods coevolved with this people over thousands of years. These people still know who they are because their ancestors and those ancestors' corn still occupy a central place in their lives.

When I returned to Mexico City from the highlands, the newspapers were abuzz with editorials denouncing American corn. Cuisine is not a trivial matter, and editorialists, in Mexico at least, understand this. NAFTA is part of a long series of devices, ranging from the padded horse collar through smallpox and the musket to the multinational corporation, for spreading the dominant culture. The treaty is helping put small Mexican farmers out of business, condemning with them their forty varieties of corn. This variety is replaced in Mexican tortilla factories with the bland American corn bred primarily for feeding livestock, and now Mexicans. It is a bitter pill to swallow. Some won't swallow it, in Mexico and in like pockets around the world. They are the heroes of this story. They are insisting on their cuisine and, accordingly, staking a claim to their humanity.

It seems to me, then, that I guard my smoked pheasants, old guitars, and quiet for the very same reasons these Mexicans guard their corn.

HOG HEAVEN

Food, like sex, is deeply personal. Unlike sex, however, it is part of public ritual and responsibility. We eat publicly but make love privately. The ways in which we copulate are, by and large, without public consequence, yet how and what we eat is an important force in shaping and reshaping culture, as it shapes and reshapes our individual bodies and experiences. Food tells a people's collective story in the same way that the molecules we eat assemble the body. The air we breathe and the water we drink (with luck and care, in the best of circumstances) are more or less the same, but food is endlessly varied. Its particularities tell much about our place in the world, just as it carries information about the changing nature of the world itself. Food is the primary incursion of the physical world into the individual.

Given what we have seen of agriculture—how it has pushed past the boundary of the planet's arable land in so many directions—is it any surprise to find that it has overrun our bodies as well? How could it be otherwise? As we have seen, industrial agriculture grows commodities, not food. The fact that commodities do not meet our bodies' needs is irrelevant.

Throughout much of agricultural history, but especially during the past forty years, agriculture has evolved in a course independent of human needs, as a force of its own, with a will of its own. Modern agriculture does not exist to serve human demands. The writer Robert Pirsig points out that to a hog in a pen it must appear that he has enslaved the farmer. Why else would the guy show up twice a day with a bucket full of feed? The hog believes this until the day he dies.

Coincident with the rise of industrialism, people started to see food less as the connection between one's body and the natural world and more as a barrier between humans and the imagined savagery of the natural world. This had more to do with industrialism's demonization of nature than with a new view of food itself, but this shift is not just academic. It changed what we eat.

To accommodate this shift, food would have to be processed—refined, purified, distilled, transfigured—into something other than its natural form. We can perhaps mark the beginning of this process with the work of Sylvester Graham, a Presbyterian minister who stumped the United States' eastern seaboard in the first half of the nineteenth century, proselytizing for what we would today call a fad food. Graham's enemy was nothing more than indigestion, and he blamed this then-feared affliction on improper diet (reasonable enough) and sexuality. Graham advised people to avoid highly seasoned foods on the grounds that spices stimulated carnal appetites. Sexual activities, according to Graham, provoked the inflammation that ruined one's physical and spiritual health. His prescription? Vegetarianism, a link between puritanism and herbivory that stretches to our time. I mention him here, however, by way of introducing a particular food he invented and touted as the key to health, the Graham Cracker. Indeed, he was on to something here, in that his cracker, different from the commercial variety that survives today, represented a return to coarse, high-fiber flours that even by Graham's time had been replaced by refined flours. He reasoned that the refined flours were too dense a package of nutrition, and he was right.

However, the future of processed foods from Graham on would follow exactly the opposite course, and all that would survive of his legacy was the idea that a food might be invented that would, in its waferlike purity, contain salvation. In this sense the line from Graham to modern processed foods is direct and unbroken. In 1863, James C. Jackson invented a breakfast food—made from Graham's flour—that he called "granula." The product was pretty awful and had to be reconstituted in milk overnight to be palatable to all but the most self-flagellating of puritans, but it set wheels in motion. A Seventh-Day Adventist and acolyte of Graham's, Dr. John Harvey Kellogg, followed through with his own cereal, something he called "granola," which he served in his sanitarium at Battle Creek, Michigan. When Jackson sued him for patent infringement, Kellogg moved on to wheat flakes, then to corn and rice cereals. Right behind Kellogg came the advertising genius Charles W. Post with his own line of processed grain cereals.

The spirit of all of this invention gets satirized in T. Coraghessan Boyle's novel *The Road to Wellville*, and Boyle had ample material. The Battle Creek, Michigan, phalanx of the movement linked up with some contemporaneous German theorists in an unholy alliance of Prussian and Puritan sensibility that spawned sanitariums and vegetarian clubs promising to purify the bodies (and capture some of the income stream) of industrialism's ascendant bourgeoisie. The movement completed industrialism's triumph over all things natural by conquering food, which was to be reduced to something unrecognizable either by machine or by an individual's brute force. British prime minister William E. Gladstone spearheaded the movement to chew each bite thirty-two times, once for each tooth, as a sort of mathematical symmetry (as if English food were not already sufficiently removed from the natural). An American businessman, Horace Fletcher, gave his name to the extreme oral processing of food, "fletcherization." Under this doctrine, thirty-two chomps was regarded as the barest minimum of mastication. Fletcher once testified that a particularly stubborn shallot he encountered required 720 jaw strokes to be brought to heel.

The point of all of this was elimination, that is, to get that nasty bit of nature through the body as rapidly as possible. The Victorians were mightily concerned with the motion of the bowels. Kellogg considered three daily sit-downs to be a mark of good health. His parallel concern reveals itself in his preoccupation with the digestive systems of young boys, for whom he prescribed enemas and mineral oil, not only to promote elimination but to prevent masturbation.

Fletcher, meanwhile, boasted that his oral rigor gave him odor-free stools, a claim he offered to substantiate by mailing samples to anyone insisting on empirical evidence.

Any of us who has been much exposed to modern food faddists will find more than campy nineteenth-century fun in this bit of history. As Boyle accurately portrays, the faddists cloaked themselves in pseudoscience, attempting to participate in the main business of legitimate science in that era: perfecting nature. In the late nineteenth and early twentieth century, diet became science. Home economists, extension agents, and chemists set about cleaning up the mess that was nature. These scientists quickly identified unnatural diets along ethnic lines, citing the cuisines of Italian, Slav, and Jewish immigrants to the United States as particularly egregious. Further, they scientifically determined that an efficient and healthy diet consisted of the very dishes—boiled, baked, and bland—that could be found right on their very own tables in WASP New England. The proselytizers devised entire cookbooks of scientific cuisine.

The food writer Jeffrey Pilcher reports that immigrants remained completely resistant to this evangelism, preferring their pastas, red sauces, horseradish, and chilies. The ascendant white middle class, however, provided ready converts in its housewives, who were the real targets of this movement anyway. Newly recruited to the science of food, they were attuned to claims of "new and improved." In this way, taste was commercialized and industrialized.

Betty Fussell characterizes this period as the time of our learning to speak "the language of commodities," the language of corn in particular. Fussell's own family was, like Graham's, Presbyterian, and in *The Story of Corn* she captures the fervor with which they adopted this new ethic:

> Infused with the spirit of the brand-name age, my family shouted hallelujah in daily worship of the processed box, bottle, and can. Presbyters who believed that wheat thins and grape juice could be transubstantiated into His Body and Blood could believe anything, anything but the equation of the "natural" with the "good." Fallen nature required redemp-

tion, and Christ was the soul's processor. We praised the manufacturers who, in imitation of Him, worked miracles of transformation. Instead of turning water into wine, they turned corn into starch, sugar, and oil.

The proto-processors focused their efforts on commodities—that is to say, grains—even before people came to think of them as commodities. No one took, say, carrots or tangerines or broccoli to a mill and attempted alchemical transformation. Chemically, those items are too complex to lend themselves to reconstitution. They are food. Yet from the very beginnings of agriculture, grain, even fresh off the stalk, was not yet food. It wanted fermenting or grinding or baking. Hunter-gatherers could be nomads because they could pluck nourishment straight from limb or bone. Grain-based nutrition required sedentism as much to process grain as to grow it. It never was immediately food but, rather, a raw material.

Not all grains, however, are equal in this regard. Wheat and rice—rice especially—are relatively close to being useful in their natural state. Rice requires only hulling and cooking; wheat only hulling, grinding, and cooking. To a degree, this is also true of some forms of corn, as traditional Central and South American agriculture knew and still knows. Yet these same people processed corn, even during agriculture's ancient incarnations.

Central to Aztec cuisine was the nixtamalization of corn, a trick that modern food processing still does not fully exploit. We know nixtamalized corn as the hominy grits of traditional Southern cooking, but its history is far older. The Aztecs knew enough to process corn by first soaking it in, then cooking it with, a solution of lime or wood ashes. These steps removed the tough outer covering of the corn kernel, but also converted some of the corn's vegetable protein to niacin and tryptophan, forms of protein more readily usable by humans. When corn became a staple of the European peasantry, economic downturns were marked by outbreaks of the diseases pellagra and kwashiorkor, both the result of protein deficiencies. This occurred because, in hard times, the peasants lived on corn alone. In the New World, the poor among the Aztecs had spent centuries living on this diet, yet remained healthier, largely because of nixtamalization, but also because their cuisine always mixed maize and beans, thus providing complementary forms of protein.

The Europeans did not adopt the Aztec process because they thought its sole purpose was to remove the skin of the corn, a task their powerful mills could handle just fine. Still, from the beginning, people processed corn, and that, coupled with the surpluses of corn that would accrue, steered the crop toward commodification as a raw material for factories.

The primary ingredient of grain (and, for that matter, of potatoes, now the world's fourth-most-important crop) is starch, which is complex carbohydrates. It is fuel. Separating that starch from fiber, germ, and hull is a relatively easy matter. The ancient Chinese and Egyptians could extract starch from rice. American colonists imported wheat starch from Europe to powder their wigs. In 1840, an Englishman named Orlando Jones invented a process that, like nixtamalization, used an alkaline catalyst to extract the starch. Colgate & Company first applied the process to wheat in the United States, but an employee saw the potential for corn and set up his own cornstarch factory in Oswego, New York. The fine, white powder spread through the culinary world about as rapidly as another fine, white powder would spread through the late-twentieth-century drug world. It was the white stuff, a hyper-refined, pure-as-snow additive for a refined, middle-class diet. Marketers even then began rewriting recipe books with cornstarch as a main character.

Even in this embryonic form, the corn-processing industry bred oligarchy. The first of these oligarchs was a marketer, Gene Staley, who established his Corn Products Company in 1906, and helped to spawn today's corn and soybean conglomerates. But an additional development was needed for conglomeration. Starch is made of complex carbohydrates; sugar is simple carbohydrates. That is, one could theoretically derive sugar by breaking down starch, obviating the need for sugarcane and the dependence on the trade with tropical agriculture that it implied. In theory, but not yet in practice.

Corn syrup was used as a substitute for molasses as early as 1733, after the invention of a relatively simple process for converting cornstarch to dextrose. By the time of the Civil War, the manufacture of corn sugar was a common process, but production really didn't take off until chemists took the next step and learned to convert corn to fructose, a sugar much sweeter than dextrose. The initials ubiquitous today on processed-food labels, HFCS, stand for "high fructose corn syrup," which was first commercialized in 1967 by the Clinton Corn Processing Company of Clinton, Iowa. Then ADM picked up the ball and ran with it.

The commercialization of corn syrup completed the commodification of corn. Each kernel was now a raw material to be disassembled and fed to separate output streams. The yellow skin and other parts make vitamin supplements (necessary now because our food is processed), but especially animal feeds. The kernel gets separated from the germ (the actual seed) and is processed to cornstarch or sugar. The germ is squeezed for its oil. Oil, starch, and sugar became the triumvirate of the Corn Products Refining Company, the brainchild of a marketer who would use these three to rewrite the design of America cuisine, first by branding it. The company gave us Mazola Oil, Karo Syrup, and Kingsford's Cornstarch. Company flacks wrote cookbooks based on these products and sold cooks on the advantages of products "untouched by human hands" in the new antiseptic factories. The starch, syrup, and oil became the basis for Bisquick, Aunt Jemima Pancake Mix, and a slew of other "convenience" products. Because Karo was a syrup, as opposed to granular sugar, and came in a clear as well as a dark form, it lent itself to all sorts of unique company-invented concoctions, such as taffy, fudge, and "divinity."

This collection of processes was in place by the time America developed its fascination with, to use a slogan of the time, "Better living through chemistry." (Some would argue that the fascination developed as a result of these technologies.) The manufactured, the refined, the convenient all became symbols of status. Until these products came along, ready-prepared foods were the purview of the rich, the people glimpsed in movies who had maids and cooks. With processed foods, any middle-class housewife could put Aunt Jemima to work in her kitchen.

The phenomenon of eating Jell-O, for instance, is difficult to explain. Jell-O is a tasteless blob of reconstituted cow's hooves artificially colored, sweetened, and flavored, served in its most revered form with lumps of corn syrup called marshmallows. Even more difficult to explain is that in the Midwest, Jell-O became a status dish, the sort of offering a beaming farm wife would bring to a church social. When I was a child in the Midwest, a row of fake-copper Jell-O molds hung in most kitchens like a collection of family crests. All difficult to understand, until a midwesterner older than I explained to me the origin of this status. To make Jell-O, one needed a refrigerator, something not at all common in the generation before mine. Taking Jell-O to the church social was a way of publicly announcing that your family could afford a refrigerator.

The food processors were not offering nutrition; they were offering the illusion of wealth, stability, and order, and consumers became willing accomplices in the plot. (It is worth remembering that the corporate marketers were simply occupying a niche, taking advantage of cheap surplus commodities to turn a profit. Agriculture itself created the niche. The marketers would have been equally happy selling most anything else, as later conglomerations of these same corporate entities would demonstrate. For instance, Philip Morris, the tobacco company, is now the owner of many of the food companies that pioneered this era of processed foods in the immediate postwar years.) Conning cooks into using up America's corn crop, however, was an uphill fight. The effort to dispose of agriculture's surplus in the human body predates the golden age of processed foods.

With the 1920s came the flapper era, and thin women became fashionable, a fact that alarmed the U.S. government. The Bureau of Home Economics of the United States Department of Agriculture, by then already a fully equipped propaganda machine for the "scientific diet," began urging women via home economists to eat more wheat—attempting to persuade them, in food historian Harvey Levenstein's words, "to start chomping their way through the wheat surplus." Indeed, this effort characterizes the USDA's course throughout the twentieth century. The agency formally had two charges: expanding markets for farm products and attending to nutrition. These roles were at odds with each other, because increasingly in the United States "farm products" meant surplus commodities—wheat and corn—and consumption of large amounts of these subverts nutrition. Repeatedly, the USDA settled this conflict by ignoring nutrition.

Women were not so amenable to the USDA's suggestions, however, even when hard times followed the flapper years. Middle-class women, at least, still ducked their duty of chomping through the wheat surplus. Oddly (or not so oddly, if one begins to grasp a history of contradictory relationships between affluence and food), the Great Depression, with its highly publicized bread lines, was also an era of dieting, comparable to our own. With women refusing to cooperate, government turned to the poor, and in not very subtle ways.

Franklin Roosevelt's initial efforts to feed the nation's hungry, then called "relief," hit a brick wall with conservatives, among them the nation's remaining farmers, who were themselves by then the recipients of multimillion-dollar federal payments designed to support farm prices. A spokesman for Indiana farmers, for instance, framed their opposition to relief for the hungry in the early 1930s by claiming the poor would spend the money on "cigarettes, malt, and other non-necessitous things." About the same time, the government bought and slaughtered six million young pigs and dumped them in the Mississippi River. It bought milk and poured it on roadways. The farm lobby that benefited from these purchases opposed giving the food to the hungry on the grounds that it would undermine markets.

When this began to produce an outcry from a public by then familiar with the sight of gaunt children and bread lines, Roosevelt established the Federal Surplus Relief Corporation in late 1933. Farmers, believing it would gut their markets, fought the new program and eventually gutted it. It reemerged, but only after the word "relief" had disappeared. The new agency became the Federal Surplus Commodities Corporation, a name that left no doubt of its primary mission, and was placed firmly in the control of the USDA. The government had overcome farmers' opposition to feeding the poor by giving the program a new goal: disposing of surplus grain. The program evolved rather early into a system of food coupons in service to its conflicting goals. Participants got two kinds of coupons; the first set, geared to wheat and other commodities, had to be used before another set for food like vegetables and fruit, not surplus commodities, became valid. Disposing of surplus farm products would remain the system's paramount goal until the Nixon administration and the federal food stamp program gave the poor the power to buy a range of food products, not just surplus commodities.

Once, with all the certainty a college freshman's knowledge confers, I offered in conversation with my maternal grandfather the mildest sort of criticism of Franklin Roosevelt. I got a firestorm response from a man who had weathered the Depression feeding his five children sometimes as an autoworker, sometimes as a trapper, and sometimes on relief. People who knew the real poverty of the Depression saw FDR as a saint. I remember clearly what those surplus commodities packages looked like into the sixties, how the boxes would come home from the distribution centers and be repackaged or put away in a cupboard, out of sight.

His wife, my grandmother, was fat and diabetic. Late in his life, my grandfather also became diabetic. Even then it was not usual for poor people to be obese from a cheap, starchy diet. In 1969, the year I graduated from high school, *The New York Times* featured photos of fat people applying for food stamps, a paradox most Americans couldn't swallow then and still can't.

Our ability to dispose of surplus, and farming's ability to extend its footprint into our bodies, expanded and matured, aided directly and indirectly by the government. The foreign front opened with a law called PL 480, which was passed in 1954 and still exists. It allows for the dumping of grain in the developing world, ostensibly as a relief effort; but as has been demonstrated time and again, this cheap grain serves as a burden to development, not only bankrupting local farmers but providing a resource for parasitical governments. By the time our government was pushed into beginning reforms of these hamhanded methods, the business of surplus disposal had been greatly refined. The poor still bore the brunt of the effort, but a range of marketing tools allowed some expansion into the general population, a necessary move as the stock of poor began to decrease in the

United States. Part of what allowed this expansion was an evolution of those fad diets of the thirties.

The first half of the twentieth century featured a rise of progressivism and faith in science. We and our leaders believed that humankind was perfectible, and science gave us the tools for perfecting it. That perfectibility, of course, extended to our bodies, so opportunity beckoned for the science of nutrition. Food faddists like Kellogg, Graham, and Fletcher begat a new and more respectable generation, many of them the products of new home-economics departments at land grant colleges. No longer was eating taken for granted, nor was enjoyment a primary goal. Eating became serious business.

In the period before World War II, Americans began paying attention to vitamins, especially the B vitamins, after Dr. Russell Wilder of the Mayo Clinic completed some cursory experiments that suggested benefits. B vitamins took on the status of a miracle drug, hailed as, literally, a "pep pill." Similarly, the term "balanced diet" came into popular usage, and the USDA issued a series of pronouncements, continuing to this day, prescribing proper intake levels of carbohydrates, protein, and fat (in ratios weighted, then and now, toward whatever happens to be in surplus). In the wake of these developments, America went into World War II every bit as food obsessed as the most committed health-food eaters of today. From the beginning, nutrition was consciously and conspicuously a part of the war effort, including a high-profile federal program for "nutrition defense." Nutritionists saw their role as an advancement of a similar effort from World War I, when the official slogan had been: "Food Will Win the War." A better slogan for the new enlightened age, the experts decided, was "Vitamins Will Win the War." Almost immediately, food producers tapped into the effort. The meat industry, for instance, mobilized both pollsters and advertising to convince people that meat should be the centerpiece of a well-balanced meal.

The federal Nutrition Division devised a seal of approval, a drawing of Uncle Sam in profile downing a forkful of food with the slogan "US Needs US Strong, Eat Nutritional Food." Producers and processors flooded the agency with requests to display the logo on their packaging, and most were successful, although only after some thoroughly bureaucratic embellishments. The Doughnut Corporation of America, for instance, was denied use of the logo if it used the term "Enriched Donuts" but allowed to use the logo when it called its product "Enriched Flour Donuts." With this sleight of hand, doughnuts became healthy enough to go to war, just as ketchup would become a vegetable during the Reagan administration. The net result of all this maneuvering was that commercial interests quickly became bound up in the "science" of nutrition, and with them came advertising, food fads, hype, and profound changes in American eating habits.

Dietary change was most conspicuous among people in the military, the impetus for all this talk of nutrition in the first place. Early in the war, draft boards found themselves rejecting 40 percent of their candidates, 33 percent of those for health reasons such as tooth decay. Justified or not, diet caught the blame for this, and the military responded with perhaps the best diet a soldier ever faced. It was abundant and hypercharged with nutritionists' favored foods of the day, such as milk and meat. More important, though, this new diet was being uniformly administered, making it arguably the biggest melting pot experiment of the twentieth century. The millions of men and women in the services, most still in their formative years, came from a variety of ethnic and economic backgrounds, yet probably ate better than they ever had before, and it was a diet formally backed by the government as the soundest science could devise. The habit stuck; the abundant, high-protein diet would become the postwar standard.

Meanwhile, on the home front, similar changes were afoot, understood best against the backdrop of the Great Depression. There were constant debates throughout the Depression about the degree of malnutrition in the country, but evidence suggests the fears were overstated. There was hunger, but most of it was confined to the poorest economic groups, which had always been hungry. The majority of people had enough food, though it came from lower on the food pyramid than during flush times. As the war heated up the economy, however, rising income was quickly converted to better food. Meat consumption rose sharply in the immediate prewar years.

Scratch any person's memory of the war years, and you are sure to hear stories about rationing—of tires and gasoline, but especially of food. Having lived through rationing sticks as a sort of badge of honor, and the two foods most often mentioned in this context are sugar and meat. Yet these memories usually fail in some details. The usual beef ration was about two and a half pounds per person per week, which by current standards is a glut. Poultry and fish weren't rationed so no one had to scrimp on protein. The average British person at the time got a pound of meat a week; most of the rest of Europe went without.

The more curious development was the war's effects on Americans' attitudes toward sugar. Although a variety of foods were rationed on and off throughout the war, most foods were not in short supply. The rationing largely stemmed from rumors, black markets, and the resultant panic buying and hoarding. The declaration of war alone brought almost overnight a nationwide run on sugar, and only sugar. People rushed to stores and bought hundred-pound sacks of the stuff. Rumors and black markets amplified the effect, and so sugar became the first commodity rationed, at a half-pound per person per week. From a nutritional standpoint, this is about a half-pound more than a person needs, and certainly enough to, for instance, cause the rampant tooth decay recorded as the reason for many draftees' rejections. Honey, molasses, and syrups were still available. Yet this bit of rationing looms in surviving memories as deprivation.

All of this points to a pivotal shift in America's food habits. Justified or not, we came out of the war craving sugar and meat, but also with the rapidly and almost universally expanding incomes to command those foods. More subtly but no less importantly, we came out with a strengthened belief that eating this and eating that could enhance not only our well-being but also our status. We accepted the notion that food could be improved. We came out of the war with a consensus as to what constituted the good life, especially with regard to food, and we came out with the tools to get it.

Immediately after the war, Europe and Asia, including our allies, faced a good old-fashioned famine, a simple lack of bread. Indeed, there was starvation. Our government knew this and prevailed upon the farmers it had been subsidizing through the years to channel surplus grain to Europe. Grain growers and Americans in general responded by not so politely telling the government to go to hell. There was more profit to be made in selling wheat to livestock growers to fatten the beef that the Greatest Generation had come to see as its birthright, just as there was intense political pressure to keep the meat supply high.

Somewhere during the mid-century ascendancy of the consuming generation, a nasty realization must have occurred to those in the food business—a realization slowly percolating since Karo Syrup's advent. Mass consumption is wholly dependent on mass advertising and, in turn, on branding. How does one, say, brand a pear as Acme's Peerless Pears when Acme's pears are pretty much the same as, say, the Polly's New and Improved Peerless Pears being offered by the competition? Evolution, breeding, and farmers make a pear what it is and not much can be added by a label.

(There is, of course, an exception to this, but it turns out to be the elusive rule-proving exception. United Fruit Company did indeed brand bananas with the label Chiquita, and did use advertising to change habits of consumption. The skim milk and banana diet invented and hyped by United Fruit was the nation's most popular fad diet of 1934, a poll showed. United Fruit, however, had to go to the trouble of cornering the market, which involved dotting Central America with its company-owned and -operated banana plantations, and taking over a government or two in the process, not the sort of thing Acme or Polly could accomplish with pears. Pears are grown in the temperate zone, not in banana republics.)

Beyond the issue of differentiating and branding, there was a second inherent problem. Demand for food is what economists call "inelastic," meaning that as incomes increase, people don't spend much more of that extra income on food. As incomes increased in the early war years, people did indeed spend more on food, buying more meat, fruits, and other quality items, but this, too, had a limit. Once the consuming public worked its way up the pyramid to meat, there was nowhere else to go. In any event, the extra money spent was not going to food processors.

The solution the food processors devised to both branding and in-elasticity was value-added manufacture. That is, instead of food, they began selling services, in the guise of processed food. People might not be able to buy much more food, but they could buy more con-venience, and layered just beneath the pitch for convenience was the message that they could buy status. Harvey Levenstein summarizes the era's promise of upper-class leisure for middle-class women:

> An executive of the American Can Company told the as-semblage [the 1962 Grocery Manufacturers Association con-vention] that "the package revolution" had helped give the American family not more time for women to work but "more time for cultural and community activities." Charles Mor-timer, head of General Foods, boasted that "built-in chef ser-vice" had now been added to "built-in maid service" implying that housewives could now lead the lives of the leisured upper class. Even in 1969, when it had become the norm for married women to work, the chairman of the board of Corn Products Company saw the "social revolution" convenience foods had brought only in terms of the full-time housewife. "We—that is the food industry—have given her the gift of time," he said, "which she may reinvest in bridge, canasta, garden club, and other perhaps more soul-satisfying pursuits."

The revolution the executive was commenting on here was a done deal in 1962, toward the end of an era Levenstein calls "The Golden Age of Food Processing," and what another food analyst calls "the Velveeta cocoon." This chemically assisted better living greatly de-pended on groundwork already laid, some of it quite deeply. The problem of differentiation, the Peerless Pear dilemma, could be partly solved by packaging, and the groundwork for that trick is as old as human evolution. Specifically, manufacturers played heavily with the colors of packaging. After one study (this process was nothing if not scientific) found that the only cars that could be distinguished from atop the Empire State Building were painted two colors, two-tone packages became the norm. Additional studies found that women responded to red, and men to blue, so package ink reflected this knowledge. In all of this, though, even if they were unaware of it, processors were tapping into that deep link between color and survival that was the basis of hunter-gatherer expertise. Using red to signal good food was about as subtle as selling *Playboy* with bare breasts.

Just as important, though, the processors relied on government-laid groundwork. Their biggest asset was the surplus of commodities, especially corn and wheat, which were simply raw energy, a blank culinary slate or platform. Properly reduced, then fortified with flavorings, preservatives, sweeteners, and packaging, these commodities could become anything manufacturers wished them to be. The public was well primed; the government had spent two generations convincing them of the value of manipulated food, telling them that flour must be enriched with vitamins, that chemical magic could build strong bodies twelve different ways and win wars. Our nation was already populated with, if not food faddists, food fetishists ripe for the picking. And the ups and downs of Depression and wartime rationing and postwar boom, of want followed by plenty, had produced a generation that would eat nearly anything if enough sugar were added. Not that the manufacturers were limited to sugar; by 1958, an inventory found 709 synthetic chemicals commonly used in food processing.

The strategy worked. The value added to food by manufacturing increased 45 percent between 1939 and 1954. Almost all of the increased spending on food during the fifties went toward manufacturing costs.

The writer Susan Allport tells of meeting a Hadza man, a hunter-gatherer, while she was traveling in Africa. "I was immediately struck by his direct manner, his easy good humor, and his black teeth," she says. Allport was looking at modern evidence of one quirk of hunter-gatherers: if they can locate bee trees, honey becomes an obsession. Allport asked the man about his favorite foods, which he listed as meat and honey. He reported drinking three mugs of honey a day.

Honey and other sugars are the wild cards in diet. Hunter-gatherers hear a sort of internal chant directing them to find more food, because within a normal range of forage, it takes all the food one can scare up to satisfy one's hunger. All bets are off with sugar, the refined essence of food, pure energy that answers that internal chant. Evolution does not equip us to deal with abundance; we have to learn to do that ourselves. The food writer Claude Fischler notes that "Societies of abundance are tormented by the necessity to regulate feeding. . . . They are at one and the same time impassioned over cuisine and obsessed with dieting."

Marketers' energies during the postwar years were directed toward the middle class for the same reason that Willie Sutton chose to rob banks. Yet if one has the tools to make commodities into whatever one wishes, then those tools can work just as well when incomes fall as when they rise, especially when sugar and corn are cheap. An extreme manifestation of this notion emerged in the sixties when food scientists began hatching grand schemes to create simple, chemically aided food for the Third World masses. Scientists touted a product they named Incaparina, a stir-and-serve powder made of cottonseed, corn, and sorghum that when mixed with sugar provided nutrition close to that of a glass of milk. UNICEF came up with Saridele, a soybean extract for infants. Chemists devised a fishmeal flour made of "trash fish" and set up a plant in Chile. The American delegate to the UN passed out chocolate-chip cookies made with the stuff. International Telephone and Telegraph devised Astrofood, a highly sweetened cupcake that made its way into school lunch programs in the United States in 1967.

All of these concoctions sound much like the livestock feed being formulated at the time. Indeed, the attitude has been that if we can concoct some form of basic gruel (preferably one that chews up surpluses) to maintain a stock of poor people sufficient to provide cheap labor and a stock of hogs and cattle sufficient to meet the culinary needs of the better folks (and somehow show a profit in the bargain), then the problem of world nutrition will have been solved. At least this approach might solve the problem of foreign hunger. The problem of feeding the domestic poor would require a bit more finesse.

As the golden age of food processing peaked, the problem of what to do about the poor suddenly became more acute, albeit not well understood. The poor, or at least the poorer, at that point were not the only folks standing in food-stamp lines. There were also people with jobs. In 1973, the real wage of American workers peaked. It has declined steadily from then until now. Ironically, this was an era in which the upper classes emerged from the Velveeta cocoon: cookbook sales soared, restaurants and diets proliferated, and tastes went global. A better-educated and rebellious bunch of baby boomers had begun insisting on something more than steak, potatoes, and Jell-O. The well-off and well-educated demanded better food, and they got it.

Everybody else got fast food and sugar.

The growth of the fast-food industry probably needs no documentation, especially not in a book. (Most Americans are far more likely to encounter the former than the latter.) Most people have seen the outskirts and now centers of every village and town spired with golden arches and the attendant cookie-cutter architecture. Ninety-six percent of America's schoolchildren can recognize Ronald McDonald; only Santa Claus surpasses him in name recognition. Internationally, the clown scores 80 percent among children. A study in the United States, the United Kingdom, Germany, Australia, India, and Japan said 54 percent of the respondents could identify the Christian cross; 88 percent, the Golden Arches.

In simple numbers, Americans spent $6 billion on fast food in 1970 and more than $110 billion in 2000. This latter figure exceeds what Americans spend annually on higher education, personal computers, or cars. It's also more than they spend on movies, books, magazines, newspapers, videos, and recorded music combined.

As we saw with the development of fish-and-chips in England, the advent of fast food does not represent a sudden break with the past, especially in light of the social overlay; fast food or something like it—largely filled with whatever happens to be in surplus—has long been the food of the working class. Potatoes are the mainstay of most fast-food operations, as they were when fish-and-chips took over the feeding of working-class Britain. The highly touted beef of fast food is actually almost waste beef; hamburger is made from the very poorest cuts left over after the steaks and such have gone to upscale buyers. (Even McDonald's fries relied on waste beef—suet—until health concerns and public awareness stymied the practice. Then McDonald's switched to vegetable oil—but added a synthetic flavor designed to reproduce the aura of beef fat.)

If Ray Kroc had a genius, it was not for invention but for spotting existing trends and tailoring his business, especially his marketing, to feed on them. Food processors had already spent a generation marketing the status and convenience of their products before Kroc came along; one can hear clear echoes of the maid-in-a-box strategy in McDonald's philosophy. "Working-class families could finally afford to feed their kids restaurant food," says John F. Love, a company historian.

McDonald's and all the rest of the chains had a clear reason for being interested in the "working class," and not just as a market. The central strategy of the chains is to dumb down every operation, to process food and freeze it at centralized factories so that it can be thawed and served at restaurants by unskilled labor. As a consequence, there are no unionized McDonald's, and the average wage of fast-food workers is now the lowest of any sector in the United States, including migrant farm workers.

That control of labor extends to the farm. The chains so dominated their markets that they could control the means of production. Farmers were drawn into "forward" contracts in which processors supplied seed and then dictated how it would be grown, when it would be harvested, and, of course, how much it would cost. The result has been a strong vertical integration and corporatization of farms. Using its buying power, the fast-food industry prompted a wave of potato-price crashes that reworked the social landscape of potato farming in areas like southern Idaho's Snake River plain. During the past twenty-five years, the number of potato farmers in Idaho has been cut in half while the total area devoted to growing potatoes has increased, as has the number of corporate-owned farms. In his book *Fast Food Nation*, Eric Schlosser reports that "The patterns of land ownership in the American West more and more resemble those of rural England. 'We've come full circle,' says Paul Patterson [an agricultural economist]. 'You increasingly find two classes of people in rural Idaho; the people who run the farms and the people who own them.' "

Similar market penetration into the beef industry has similarly reshaped its social structure from farm to fork. The most telling of the many statistics one could site is that the suicide rate among American farmers and ranchers is three times the national average.

The fast-food industry also worked a field already plowed by the earlier generation, advertising—particularly television. TV not only transmitted the new culture, it became the culture, advertising fast-food products and serving up prepackaged entertainment to an audience that watched as passively as it ate.

In its essence, though, the fast-food trend is sugar. Just as it tapped changes in the labor market, fast food's rise corresponded directly to the food-processing industry's, especially, as we have seen, Archer Daniels Midland's decision in the seventies to go whole hog into the business of converting corn to sugar. Cheap corn made cheap sugar, which became the common denominator of fast food, appearing in everything from "special sauces" to hamburger rolls. As the trend has evolved, and as consumers have been trained, though, fast-food firms have become less subtle, putting only the merest façade of food on its sugar. Now sugar needs only the addition of water and a bit of coloring and flavoring to become marketable: soda is the high-profit item of the fast-food business, a fact that has also benefited convenience stores. Even more widespread than fast-food joints, they have stripped away the veneer of food altogether, going straight for the high-profit "big gulp." A thirty-two-ounce soda and a tank of gas is America distilled to its seminal fluids.

Not that fast-food sellers are standing on the sidelines. They are capitalizing heavily on a hole opened by the soda manufacturers, specifically Coca-Cola, Pepsi, and Cadbury-Schweppes (which makes Dr Pepper). These makers figured out that the expansion of their business (their goal is 25 percent per year) simply could not be maintained by a market of adults only, so they began a campaign to reach children, principally through television, but also, and more insidiously, in the schools. By contracting with school districts, soda manufacturers get schools to host soda machines in exchange for a cut of the profit, thereby giving them a financial stake in increased sugar consumption by the kids they are supposed to educate and protect. It works.

Meanwhile, at fast-food places, larger and larger sodas have become a marketing gimmick and profit booster. In the 1950s, a customer got an eight-ounce soda with his fast-food burger. Today, a child's meal includes a twelve-ounce soda, and a large soda is thirty-two ounces. Soft drink consumption has quadrupled in the United States over the past fifty years. Just since 1978 the intake of a typical teenage boy has tripled to more than twenty ounces per day. One-fifth of the nation's one- and two-year-olds drink soda routinely, some of it from baby bottles.

I began planning this book after a single jolting experience. I had been traveling in Peru and Chile for a couple of weeks, researching efforts to feed the poor in the developing world. It's odd how a few weeks abroad can subtly remove the salient points of America from one's view. Never did I expect to encounter culture shock on reentering my own country, but I had gotten off an all-night flight from Lima at Los Angeles and begun working my way toward a domestic connection, when I found myself overwhelmed with just such a shock. It was based on a single observation: that Americans are fat.

Obesity is pandemic in the United States, and is now officially recognized as a leading threat to public health. There is a tendency to frame this problem in economic terms, especially when one has spent time sitting in the mud huts of people who live on less than a dollar a day. Corpulence is indeed an economic artifact, but in exactly the opposite way that one might have expected: obesity in the United States is not a mark of wealth; it is more reliably a mark of poverty.

Doctors who have grappled with the problem of obesity cite a variety of causes; some even disagree on the problem's distribution among social classes. Clearly, there is more at work than simply surplus corn and sugar. There is a complex psychology to weight, so much so that in 1978 the magazine *Psychology Today* reported that food had replaced sex as Americans' leading source of guilt. We are, after all, not evolved to deal with surplus and have deep-seated drives to consume as much as possible, to lay in stores for hard times. This sets up a classic love-hate relationship, a paradox, and we can see evidence of conflict all around us. Young upper-class and middle-class women suffer in record numbers from eating disorders such as anorexia and bulimia, while obesity is rampant among other groups.

Yet those same doctors and researchers struggling with this problem tend to cite a common denominator: sugar consumption. On average, Americans today get about 16 percent of their total calories from added sweeteners, mostly corn syrup. Current dietary recommendations say we should limit that to 6 to 10 percent. The biggest source of added sweeteners is soft drinks.

A second common denominator, especially among obese children—and in that age group the problem is truly epidemic—is sedentism, or more to the point, inactivity, because fat kids spend more time than other kids watching TV.

We have seen a decline of the pernicious and ham-handed methods of the 1930s, when the poor could not get food unless they agreed to eat their way through the surplus stock of corn and wheat, because our methods of ensuring this same outcome have become more sophisticated. The methods are new, but the goal itself is as old as agriculture.

A COUNTERAGRICULTURE

This book has progressed to the point where the dutiful author, having carefully considered all the relevant points from ten thousand years of human history, is to execute a few deft strokes that will right all wrongs and move us toward, to use Christopher Lasch's appropriately sardonic label, "the true and only heaven." His book by that title is a much longer and more thorough indictment of the progressive fallacy as it played out in the United States in the twentieth century. Agriculture and rural communities were the chief targets of the progressive movement in the United States, but where they are concerned, we need not limit the span of the indictment as he did.

The notion of progress is a creation of agriculture. I can't prove it, because the idea of progress is too deeply burned into us to be extracted and examined. It is our secular religion. Nonetheless, it is easy enough to imagine those millions of years of hunter-gatherer existence when the terms on which life was engaged remained unchanged, century upon century, never mind that no one had need for the concept of "century." Hard to imagine there would be an expectation of change then, or any sort of belief that change would be a good thing, that one must better one's existence or reform the world. We had not yet imagined that we had been given responsibility for the world—"dominion," as the religious among us call it. We humans were simply of the world.

It's equally hard to imagine humans not claiming some right to control life after agriculture's advent, when we became demonstrably responsible for crops, when what we did or did not do provided abundance and reshaped the physical world. Harder still to imagine us avoiding hubris after all that mounding, storing, counting, and accumulating of grain—after power accrued. It was easy enough then, at least for the powerful, to assume that we had been placed in charge of our destiny.

I don't mean to argue here that we have no responsibilities for the future; we accepted some with agriculture, and they cannot be repacked into Pandora's box. Nor am I saying that agriculture itself was a bad idea, a mistake made by humanity. Certainly the consequences have been dire for both humanity and the planet as a whole, but agriculture was not a bad idea, because it was not an idea, not an invention, certainly not an advance, as it is so often labeled. It arose independently in at least five different places on the planet. In each case, it arose as a necessary consequence of natural forces, which is to say, it evolved. We did not create it; in a very real sense, it created us. Plants domesticated us. The distinction is not sophistry or philosophical hair-splitting, and it points out some real limitations on our options from here.

Chaos theory has gone a long way in helping us to understand emergent behavior and evolution, the ability of systems made up of senseless elements to take on a sense, a logic, rules of their own. The sum of much randomness is order. Agriculture is such a system. Some students of such theory hatched an apt analogy for the political issues of chaos: the plight of a political actor in our world is much like that of the person at the oars of a raft in white water. The analogy rings true for anyone who has had the real experience. A raging, roiling river is nothing if not chaotic, and it makes it instantly clear to any rafter that he has no say over the general course of travel. The raft's only workable course is the river's course, but the abilities of the rower still make a huge difference as to the success of the trip. A white-water pilot is an opportunist, reading the flows and eddies, the rocks and contrary currents, sometimes executing maneuvers that are counterintuitive, all in the interest of staying right side up. The river composes the theme; the raft is contrapuntal. If all this sounds terribly daunting, bear in mind that some of us run rapids for fun.

Modern reformers need to remember characters like Henry A.

Wallace. The history (and present) of agriculture is rife with all sorts of progressives. Almost everything that has been wrought, every innovation, has been hatched with good intentions such as increasing yield, feeding the poor, or protecting the family farmer. In the end, most of those innovations have had the opposite effect from what was desired, or have created a new round of more intractable problems. The agricultural system has evolved to take any energy applied against it and use it for its own purposes. Efforts to reform farm policy and its subsidies favoring big business have been mounted repeatedly for at least two generations. During every round of this fight, coalitions have agreed that current policy is on its face absurd. Still, not only does the folly continue, but it grows.

The political system cannot be counted on to reform agriculture because any political system is a creation of agriculture, a coevolved entity. The major forces that shaped and shape our world—disease, imperialism, colonialism, slavery, trade, wealth—all are a part of the culture agriculture evolved. We carry evidence of its disease in our bones and blood, of its pollutants in our cells, as surely as those ten-thousand-year-old skeletons of farmers are deformed and decayed by the very same infectious diseases, stoop labor, and exploitation. Just as surely, agriculture dug the tunnel of our vision.

Truly, there are steps to be taken, changes to be made, but the last place we can look to make those changes is at the top of the political hierarchy. That only ensures that our drive for change will be converted to currency in ADM's account.

As I acknowledged at the beginning, my attentions in this book are largely Eurocentric, toward the mainstream culture. But the mainstream is not the only stream. There has been for all of these years a subculture, a food counterculture in the largest sense of the notion. It survives in myriad and often ignored forms, and it can guide us.

Understanding the essence of the subculture requires some definitions, and some adjustments in our thought. For instance, if by agriculture I mean growing food, then I've got a problem when I assert that agriculture is a bad idea. Obviously, we need the food; hunting and gathering, commercial fishing, and the like will not meet the needs of six billion people. I do not take population as a given; if we accept that six billion as inevitable, we are doomed. If the human endeavor takes as its primary reason for being the feeding of however many people issue from senseless acts of reproduction, then the human endeavor is pointless. Nonetheless, we will need to grow food.

There is a distinction to be made between what I have called agriculture and simply growing food. Call the latter farming, a distinction that still has its problems, though it will serve for a first cut. The difference is that the goal of agriculture is not feeding people; it is the accumulation of wealth. What agriculture grows is not food but commodities, grain not to eat but to store, trade, and process. Consider the range of plants humans consume, the hundreds of species. That's food. Consider that two-thirds of our calories come from wheat, rice, and maize. Add sugar and you have a nearly complete picture of commodities. It is an oversimplification, but a useful one, to assert that these commodities have a fundamental and key distinction from the rest of food: they are storable and interchangeable and close to currency in their liquidity; in fact, they are traded in markets just as currency is. They form the basis for the accumulation of wealth, and have done so for ten thousand years.

U.S. policy has stated time and again that agriculture does not grow commodities in order to feed people. For most of the century and a half of its existence, the role of the Department of Agriculture's nutrition arm has been to get people to eat more of whatever commodity happens to be in surplus. One of the blunter statements of this conflict came in 1979 when the USDA, as it periodically does in fits of reform, began drafting a new set of dietary goals for Americans. Harvey Levenstein summarizes a pivotal development in that effort: "Bob Bergland, [President Jimmy] Carter's secretary of agriculture, even resurrected the long-spurned idea of the 1930s that nutrition and health, not selling food, should be the goal of federal farm policy—though at some future time, he cautioned, and he was out of office before that future arrived." Tying food to the health and well-being of humans is heresy within our—and all—political systems.

The markets, corrupted as they are by subsidy, have no way of dealing with health and nutrition, but the pristine markets of free-marketers' wet dreams would hardly be any better at it. As Lewenstein points out, when people became concerned with the quality of their food and fomented the health-food movement of the seventies, it took almost no time for the major food processors and corporations to begin marketing health-food brands of their own, or failing that, to begin gobbling up the small independent businesses that were feeding the health-food movement. Still we are assaulted with labels from all quarters proclaiming that this or that product contains no cholesterol or sodium or whatever happens to be the latest scare, never mind that the product is laced with sugar.

Markets cannot make the changes that must be made because of the need for what Lewenstein calls "negative nutrition." That is, to be healthy, Americans first need simply to stop eating some things: most fats, sugar. Nothing is required as a substitute, so nothing needs to be sold (the supplement industry's entreaties notwithstanding). How does one take to the airwaves with a multimillion-dollar advertising campaign urging the consumption of nothing? Who pays? The public-interest money for promoting health and nutrition is a pittance compared to advertising budgets for, say, soft drinks, and even the public-interest messages are watered down by industry lobbying.

We grow wheat, corn, rice, and sugar not because that is what people want or need, but because that's what we know how to grow well. We know how to grow these few crops well because commodities built the culture of agriculture, a spinoff of which is investments in research and development. Worldwide, the money spent on agricultural research has been spent almost exclusively on these crops. It is not a stretch to say that advances in other crops have come only because something learned about the mainline commodities proved, by chance, to be applicable to a forgotten or orphan crop.

This then presents an enormous opportunity, as some researchers have already discovered, and not just for researchers. More important, buried in this is the signpost to most of what we need to know. We have spent a great deal of time talking about what agriculture is, the acknowledged bias of the argument, but this line of thought can also give us a clear idea of what agriculture is not. By paying attention to that which is food but not agriculture, we begin to see a line through the rapids.

It sounds odd to say it, but agriculture is not about raising food. Still, if it's food we need, then what would we call a system with the clear goal, not of accumulating wealth but of the earthshakingly radical idea of attending to human nutrition, health, and well-being in the broadest sense? I am less interested in what we might call it than in what it might look like.

For openers, it would not deal with commodities. It would deal with the staggeringly long list of familiar fruits and vegetables, but also grains and pulses and the unfamiliar regional crops that agriculture orphaned and forgot. In drawing a distinction between food and commodities, I am proposing a somewhat permeable line between the two. This is not law, after all, but a thought experiment.

What I mean by food are those crops people grow to eat directly, as in the traditional distinction in agronomy between commercial or industrial agriculture and subsistence agriculture. Wheat trucked away from a three-thousand-acre farm is an example of the former; the tomatoes from a suburban Minneapolis backyard garden or a hectare's worth of sweet potatoes, bananas, and cassava growing around a grass hut near Kampala are cases of the latter. So, too, are some commercial forms of farming—that truckload of vegetables from an Illinois farm headed for a farmer's market in Chicago, for instance. The vegetables are perishable and go from farmer to consumer unprocessed.

Despite our living in an agriculturally dominated society, food is still readily available to us, even in grocery stores. To say that it is everything without a corporate logo on it may be an oversimplification, but not by much. Let's complicate it a bit. Rice is the commodity most likely to slip past this definition since well over half of all rice consumed is eaten by the same people who grew it. That sounds like subsistence agriculture, so am I asserting that this portion of the rice crop is food, and all the rest is commodity? Yes, I am. True, rice is storable, tradable, a dense package of carbohydrates that meets the definition of a commodity; but because it is the most important foodstuff for some of the world's poorest people, it also has many of the hallmarks of food. It doesn't have to be processed or packaged. More revealing, though, is that when the Rockefeller Foundation did a survey of the state of crop research in the mid-eighties, it found science was clearly neglecting rice, partly because it was a foodstuff for poor people. Science decided rice was food, not a commodity, and behaved accordingly.

The same chameleon definition applies to potatoes. For all of the reasons they swept through the backwaters of Europe two centuries ago, they are becoming one of the world's most important subsistence crops. At the same time, as we have seen, corporate control of the fast-food industry has made them behave very much like a commodity in the United States, as well as in places like Chile.

On the other side, crops that one would normally consider food, say bananas, can behave like commodities. Americans like bananas but can't grow them, which means a controlled network must grow and import them. Thus, under the old United Fruit Company, bananas spawned a plantation system analogous to the slave-dependent sugar trade that fed colonial Europe. Some of United Fruit's excesses in maintaining that network rivaled those of slavery. Clearly, though, we need a definition more linked to process than to crop in order to think about this. The practical value of this approach is that it reveals niches not filled by industrial agriculture, and that is where opportunities lie.

The biggest such niche is geographical. By and large, industrial agriculture is an endeavor of the temperate climates. We spend a lot of time speaking of and trying to define the "developed" and "developing" worlds. It's simpler than we think. The developed world is just Europe and its former temperate colonies, and Japan, itself temperate. The neo-Europes, every one of them temperate, are responsible for most agricultural exports. Industrial agriculture has tried to colonize the tropics, but has largely failed, which is why the neo-Europes are neo-Europes, while other former colonies, such as India and the countries of Central America, Southeast Asia, and tropical Africa, remain their own creatures. With the exception of expensively maintained plantations like those for sugar, bananas, and coffee, European agriculture bounced off the tropics.

As with the absence of research on orphan crops, the absence of industrial agriculture in the tropics provides real opportunity. It offers a chance to do good work out of sight of—under the radar of—mainline industrial agriculture. This is more than a theoretical advantage.

For example, much has been made of corporate control of new technologies, especially biotechnology. Corporations like Monsanto have exercised that control by asserting intellectual property rights through patents. This is a real issue in the developed world and a feared issue in the developing. In reality, though, people working on forgotten and orphan crops in the developing world are finding themselves unfettered by corporate influence; these crops do not register on the corporate radar, nor are they likely to, simply because they offer little potential for profit. In some cases, corporations like Monsanto have been more or less shamed into sharing some of their technology as it applies to developing-world crops, knowing that doing so will cost the corporation none of its market share in the developed world.

That profit-driven bias can open enormous holes and enormous potential if we simply re-ask a few key questions. By and large, U.S. agricultural research, both corporate and public, has been profit-driven. It plays around with various inputs to increase yields, simply because somebody sells inputs. Thus, hybrid seeds could develop quickly in the United States because someone could sell the seed, but more important because hybrid vigor does not pass to succeeding generations, so farmers would have to buy that seed year after year. Asking scientific questions in only this way for more than a century has created a large unexplored territory.

The agronomist Chris Mundt, for instance, was part of a group that decided to test a simple idea in Asia. It is generally understood that companion planting (or intercropping) of various species causes the phenomenon of overyielding, in which each plant produces more than it would if grown alone. However, with the row cropping and mechanical harvest of monoculture, this is not a practical system of agriculture on any sort of scale, so fertilizer substitutes for intercropping. Mundt, however, tried to obtain overyielding by planting together not different crops, but different varieties of rice, which could be uniformly harvested by machine. It worked spectacularly. This sort of research will not come out of a corporation, simply because the results don't require anyone to buy something. A farmer buys as much seed as usual, probably less fertilizer. The solution is simple, elegant, and cheap—but for suppliers, unprofitable.

Likewise, some public-interest scientists are now working with the tools of biotechnology to develop a form of asexual reproduction called apomixis, in which offspring are clones of a single parent. Were this technology developed, hybrid vigor could be passed to succeeding generations, meaning farmers could save seed. No corporation is going to develop this technology. In fact, agribusiness can be counted on to vigorously fight its development.

It is easy enough to see how research's bias toward three or four crops has left hundreds of other crops with all sorts of untapped potential, and this effect is greatly magnified when other biases of research are taken into account. What can be done with small-scale, labor-intensive, tropical, subsistence agriculture by working on perishable, orphan, or forgotten crops that are eaten by the people who grow them or sold, unprocessed, in direct markets in a nearby village or town?

What about the developed world? Much of the same potential applies. Consider, for instance, the massive public investment in industrial agriculture, including not only subsidies of thirty billion dollars in some years but also transportation networks like locks and canals, railways, highways, and fuel, not to mention the enormous subsidy of research at land grant colleges. The product of all this investment is a social system that is by any reasonable measure dysfunctional. Small-town life in the American Midwest has become untenable. Despite its stated reasons for being—to diversify agriculture, and to keep farmers on the farm—the Department of Agriculture has, since its founding in the nineteenth century, produced two unbroken curves in the opposite direction.

Nonetheless, there is still growth in agriculture in the United States—or, in the terms of this book, growth in farming. Organic agriculture is the fastest-growing segment of American agriculture, expanding at a phenomenal rate. The number of organic farmers in the United States doubled in the last half of the nineties. Organic farming is expected to claim 30 percent of the European market by 2010. This with virtually zero public investment.

Among industrial-agricultural agronomists, economists, and bureaucrats, one often hears laments of declining public investments in agricultural research (of the sort they favor). Given the record of their research and the healthy growth of organic agriculture in its absence, one wonders how much more investment in research we can tolerate. Indeed, if tomorrow the entire weight of the U.S. agriculture budget turned on a dime and directed itself away from wheat, corn, soybeans, cotton, and rice and toward sustainable organic agriculture, I suspect that the result would have nothing to do with sustainability and organic practices and everything to do with profit. Given the state of affairs of the oligarchy, the best of all possible worlds is to be ignored by it.

Consider a seemingly trivial example. A generation ago, the state of American beer was deplorable; watery, chemical-induced brews issued from the vats of a few corporations that spent most of their money on advertising, not quality. Of course, no one involved in federal policy was likely to do anything about this, even if one could imagine a federal government ready to do battle with the Coors and Busch families. The solution was a lot easier than that. People started local microbreweries, unnoticed except by a few customers in the towns where they started. They charged double the going rate for a glass of beer and people gladly paid it, and a movement spread, quietly, beneath the notice of the corporate world, until it was too late and the microbreweries had captured a huge share of the market— bit by bit, unsubsidized, unlobbied, with the simple expedient of quality.

The example is in fact not trivial, because beer is made with grain. Fred Kirschenmann, the organic farmer we met earlier, now directs the Aldo Leopold Center for Sustainable Agriculture at the University of Iowa. He reports that the emergence of microbreweries was a key factor in making organic farms profitable. There has been a parallel surge of small, artisanal bakeries, where people line up to pay maybe three times the supermarket rate to buy a loaf of bread. Simple, homely bread baked skillfully with quality ingredients.

Agricultural issues are largely ignored by Americans, a factor in the creation of the political vacuum that is the niche of corporate lobbyists. Americans dismiss these issues because they assume only rural areas are affected, and not many of us live there. The assumption, however, is false. Virtually every one of us faces the consequences of our ignorance of agriculture three times a day.

One small exception has been the attention given to a quiet but significant and encouraging development that begins on farms but directly affects cities. For example, this 2001 report in *The Christian Science Monitor*:

> It's 6:30 a.m. While most New Yorkers are still sleeping, farmers—at a sprinter's pace—are turning one of the city's urban parks into a real-life cornucopia. White-canvas shelters are popping up like mushroom caps. Folding tables snap open, and then, faster than you can say "heirloom tomatoes," the bounty is ready. There are eighteen varieties of onions, yellow carrots, striped beets that have concentric pink rings and a mild taste, and a squash that looks like a medieval trumpet—long, thin, and curling back on itself as if nature had second thoughts about this plant. Buckets hold flowers so fragrant bees arrive, looking for pollen. In fact, the bees might as well have their own stall, because not far away are jars of New York City honey. You're right. This is not the produce aisle of your local supermarket. It's a walk on the "wild side" of fruits and vegetables. This is New York's Union Square Greenmarket.

The phenomenon is not at all unique to New York. The same story reported a 63 percent increase in the number of farmers' markets nationwide between 1994 and 2000, a growth that parallels the boom in organic agriculture over the same period, one of the worst for commodities farmers. There are now 2,800 such markets spread across the United States, with 19,000 farmers selling wares. Annual sales at Seattle's farmers' markets went from $5 million to $19 million between 1997 and 2001.

Yet those gross sales figures underestimate the importance of the markets to farmers. A gross of $19 million is trivial when stacked against commodity sales, even trivial when stacked against the sales of a single megafarm. But the average commodity farmer keeps less than 5 percent of his gross. A vegetable farmer selling directly to consumers keeps most of it.

Still, dwelling on the economics of this movement misses the greater significance of these markets. Rereading the *Monitor*'s description above, one is struck by the degree to which the reporter was struck by variety. Buyers go to farmers' markets because they are presented with a range and quality of food that the existing commercial system has long denied them. Walking through a farmers' market, one is struck equally by the variety of conversation, especially between farmers and consumers. Buyers don't just buy; they learn. One farmer, for instance, is selling greens with bug-chewed holes in them, a sure sales-killer in a supermarket produce aisle. The farmer, however, is there to explain that the bug holes, besides being harmless, are the best organic seal the greens can bear; the holes mean the greens haven't been sprayed with a pesticide. Farmers are present to explain the unfamiliar, to urge consumers to try new greens like pokeweed, to explain how that weed helped people in the Midwest make it through the Great Depression. The buyer carries home some social history along with dinner.

All of this, of course, is terribly inefficient, and that is precisely the point. Far fewer man-hours and dollars are spent in mechanically picking and shipping to supermarkets perfectly uniform, bombproof vegetables. Many consumer hours are saved when people whisk through a supermarket plucking prewrapped, and usually precooked, food substitutes from shelves. Who *said* food was supposed to be efficient? Are we better served by spending the time we save in front of television sets, consuming packaged, standardized images, or by lingering amidst the lushness of a community market?

Open-air markets are the hub of most any city or town in the developing world. The reasons are far more than commercial; this is where people connect with the substance of their lives through the experiences of other people—real, physical connections. I wonder what these people in "primitive" towns would think of our having to reinvent now what they never lost, of the academics endlessly musing over the loss of community in our society when the material of community is so simple, physical, and fundamental.

Community is being quietly rebuilt in the countryside as well. America's Northeast was once U.S. agriculture's major force, with farms supplying the country's densest clusters of population. The Northeast now does very little farming, and most of the agricultural lands have gone back to forest or suburbs, an odd transition in that these are in some ways the nation's best agricultural lands. Unlike the western grasslands, the Northeast gets enough rain to grow crops. The decline came as a result of a particular form of subsidy: federal irrigation projects. Beginning at the height of the progressive era, the nation set to work on making the western deserts bloom, investing billions in dams, canals, tunnels, and drains to bring nine million additional acres under cultivation. The historian Donald Worster has shown that this figure exactly parallels the acreage of abandoned farms in the Northeast—land that already had water—during the same period. What industrial agriculture abandons is opportunity.

In the fall of 2000, *The New York Times* reported a new back-to-the-land movement in New England, a series of institutes and schools set up independently but all more or less teaching would-be farmers the skills of growing and selling vegetables to resurgent farmers' markets:

> These internships, which entail working and often living on a host farm in exchange for a stipend and the experience of a growing season, are an increasingly popular recourse for a new generation who did not grow up on farms but who are drawn to that way of life. Part of the lure, as always, is the chance to live closer to nature, but these aspiring farmers also know they need to be well versed in the business end of farming—and

see a future that makes financial sense. They are taking up a task others are abandoning . . . [but] these unseasoned farmers are hoping that a renewed interest in organic foods and locally grown produce will make their dream of running a farm of their own a reality. The New England Small Farm Institute, a nonprofit educational organization that helps coordinate internships on about 70 small farms in the region, sees everyone from recent college graduates to burnt-out business executives asking how they can learn to work the land.

Unlike the back-to-the-land scare of the sixties and seventies, however, the fruits of these efforts are not flowing simply to hippie health food stores, although many of those survived, evolved, and now form a key node of this network. Consider the *Monitor*'s report: "On the bags [at the Greenmarket] are the names of some of New York's tonier eateries: JUdson Grill, Gramercy Tavern, and Tabla. There are long green dandelion leaves, which will be used with skate and chanterelles; purple kohlrabi that will go raw into a salad; fennel that will end up with an octopus vinaigrette."

Boutique vegetables? Just another yuppie indulgence? There's another way to look at this.

As is the case with most former sixties radicals, one would not guess Alice Waters's past to look at her today. True enough, she is still in Berkeley, but her home base happens to be one of the world's most prestigious restaurants, Chez Panisse. The lunch crowd, mostly older folks and a few tourists comparing restaurant notes, lines up a good half-hour before the sliding wooden gate out front opens. Waters's assistant reminds me that she has only twenty minutes, as he will remind Waters repeatedly during a somewhat longer interview. She has to make a meeting in San Francisco. He packs a sandwich for her to eat en route.

Petite, pretty, flashing blue-gray eyes, she is all motion and activity, as one would expect a businesswoman of her stature to be. She worries that my tape recorder will catch her being inarticulate—an unjustified fear—all the while nervously filling my water glass for me as she fields each new question.

I am interested in how a bona fide sixties radical decided to pursue her politics in the restaurant business, and she tells me it wasn't nearly that direct. She began her enterprise in 1971 after a watershed trip to Europe.

"I went to France. I had a sort of sensual awakening, if you will. I'd never eaten anything good before, certainly nothing exotic, nothing tasty," she says. But the experience went beyond food. "People were buying in the market every morning, then eating with their families, then hanging out in cafés in the afternoon where they knew nearly everybody in their neighborhood. I wanted to live like that. It was comfortable and meaningful for me to be engaged in that way."

So she opened a restaurant, but at that point she had no idea that this simple act was about a lot more than food. She only knew it began with food. For the first twenty-five years, Chez Panisse was a way of serving quality. At the end of that period, though, she reflected, and it occurred to her then that something more than cooking had happened.

From the beginning, Waters decided that quality had to do with freshness, and fresh meant local. Restaurants were simply not supplied with local produce in those days, so her staff became, as she put it, "hunter-gatherers." They built a network of local organic suppliers of meat, fish, and vegetables from the ground up, a network that still feeds into Chez Panisse, not to mention the rest of the Bay Area's distinctive cuisine.

Only after twenty-five years did it become apparent that the real significance of the restaurant was not the menu causing the tourists to salivate and gasp as we talked, but the larger community Chez Panisse had assembled. "Something had changed in Berkeley in that very short period of time, something dramatic and hopeful," she says.

Accordingly, food has become her politics, although there is nothing inherently radical in catering to upscale tastes of one of the wealthiest enclaves extant. Which is why six years ago Waters extended her work to schools, with what she calls the "edible schoolyard." In this experiment, schoolkids grow the vegetables used in their own school lunch programs, a means of feeding and subverting young minds.

"Every choice you make has consequences for the quality of your own life and good health, but also for agriculture and culture. If you buy food from people who take care of the land, you are supporting a whole way of life," she says. This is not about catering to affluence so much as it is about teaching Americans a fundamental value that poor people elsewhere in the world take for granted.

"In other cultures there is a nourishing, inexpensive, delicious kind of cooking that is acceptable to everyone. That should be a right, not a privilege," she says. "I only want to have the food that is alive and ripe and delicious."

A radical thought, indeed.

In the above discussion I have been indulging in the terms "organic agriculture" and "sustainable agriculture" with the inexactness most of us allow these ideas. As we generally understand these terms, they are useful, in that they describe a form of agriculture that self-consciously steers away from the worst excesses of commodity agriculture as it has existed for ten thousand years. The fundamental attractiveness of organic agriculture is not that it provides pesticide-free food to the affluent, but that it pays attention to cycles—life cycles, and especially nutrient cycles. It consciously abandons the industrial model of input/output/waste and instead closes loops to resemble self-sustaining natural processes. If all of humanity began overnight to shop exclusively at organic farmers' markets, though, our troubles would not be over. Organic agriculture is still agriculture, in that it relies on a relatively small range of plants that are evolved to follow catastrophe and thus require disturbance and are primarily annuals. Organic agriculture is a necessary step, but it is not sufficient, at least as it stands; a fundamental redesign is required.

This redesign will require something approaching permaculture. We have some of the tools for this now; in fact, some of our food has long come from perennials like fruit and nut trees, along with a wide variety of berry bushes. We can increase reliance on these, but we need to focus on inventing a perennial polyculture, a system in which a variety of plants grow together permanently, performing services like fertilization and defense from insects for one another as well as providing food. It is a vast undertaking that requires not so much a redomestication of crop plants but rather a reinvention by selection, by breeding, by choice, by patient learning, by reexamining the genetic diversity that remains, by creating feral farming. Not back to the garden, back to the wild.

A movement began in Italy that now claims sixty thousand members worldwide. It calls itself the Slow Food Movement, a wonderfully subversive name, not so much because it counters fast food but because it celebrates the notion of going slowly; it flies in the face of efficiency. I do not rule efficiency out of my life, but it would be a sad life indeed if I let it rule. Yet efficiency does rule our food and in so doing has stripped food of its fundamental meaning. Our communion with creation has been reduced to maximizing yield per acre, calories per dollar.

Fittingly, the Slow Food Movement has a manifesto, adopted in 1989. It says:

> Our century, which began and has developed under the insignia of industrial civilization, first invented the machine and then took it as its life model. We are enslaved by speed and have all succumbed to the same insidious virus: Fast Life, which disrupts our habits, pervades the privacy of our homes, and forces us to eat Fast Foods. To be worthy of the name, *Homo sapiens* should rid himself of speed before it reduces him to a species in danger of extinction. A firm defense of quiet material pleasure is the only way to oppose the universal folly of Fast Life. May suitable doses of guaranteed sensual pleasure and slow, long-lasting enjoyment preserve us from the contagion of the multitude who mistake frenzy for efficiency.

The use of the word "sensual" is key. Much of what we need to know to free us from agriculture we knew ten thousand years ago.

The goal of attending to the nutrition of people is radical, but I don't think it radical enough. Food is about a great deal more than nutrition. It, along with sex, forms the pathway that connects our species to the future. Evolution hinges on survival, which in turn hinges on nutrition and reproduction. We have other needs, oxygen for instance, but these are automatically met. We must hunt for food and sex. This hunt is our obsession, our drive, the focus of our senses and our sensuality, so ingrained as to define our humanity. These drives are our essence.

Somewhere along the line we became so focused and so competent in this hunt that we rigged the outcome. To hunt is to be insecure about the immediate future, to experience the nagging fear of want that has driven us to our worst excesses and finest creations. Agriculture rigged this game by allowing storage and wealth, ensuring future food (and sex). Agriculture dehumanized us by satisfying the most dangerous of human impulses—the drive to ensure the security of the future. In this way we were tamed.

Yet the hunter and gatherer survives in each of us. When a woman ambles through the Union Square market and the deep purple glint of a plum catches her eye, she is replicating a primal process, awakening pathways of primal signals. The process itself is satisfying, human. When she speaks with the farmer who grew the plum, she connects to a bit of her community, her link to the rest of humanity. We subvert agriculture every time we reestablish that link. Our weapon in this is sensuality.

I EAT, THEREFORE I KILL

Consider the word "venery." It has two meanings: "the gratification of sexual desire," and "the practice or art of hunting, the chase." Its root appears throughout the Indo-European languages in a variety of permutations meaning to want, to desire, to pursue, to hunt, to lust. Think of Venus, the love goddess, whose name partakes of this lineage. Also "venereal," "venison," and "venerate."

Webster's labels both of the word's meanings "archaic," as if the conjunction of these ideas survived to the sixteenth century, then died, not to be revived in our era, sanitized as it is of base lusts. To desire, to pursue—both verbs are now reserved for consumer goods. We would like to think the conjunction of food and sex ended when our line split from the venal bartering of chimpanzees.

Years ago I used to go hunting in eastern Montana, an eight-hour drive from my home on the state's western edge. I went not so much for meat as for the excuse of walking alone for a few days in one of my favorite places on the planet. Eastern Montana is quintessential high plains, which places it on the ragged edge of agriculture. Stretches of it, especially the stretch Montanans call the "Golden Triangle," just east of Great Falls, are flat and tame enough to welcome a midwesterner. It gets enough rain (most years), has enough topsoil (or at least did), and is monotonous enough to grow wheat, and it does, usually in plots of three thousand or four thousand acres owned by a single farmer and neatly sliced into miles-long strips of wheat alternated with miles-long strips of fallow land that next year will be wheat. This is Choteau County, the place *The New York Times* profiled when it needed to make the case that farm subsidies have created a welfare state.

It is a welfare state precisely because it is an edge, the very limits, of farming. It's easy enough to cross to the other side, up coulees, dry gulches, behind buttes and foothills, into badlands where wheat won't grow and behemoth tractors won't stay upright. Through generations people have tried to make a living here, and they have failed. It's beef country largely, grazing land, but even that has its limits when the parched soils won't raise enough grass to justify cows.

Still further east, out past Havre (a name that accurately reports European designs on the place), where the Bear Paw Mountains poke up out of the plains, a county-sized span is not farmable. Naturally enough, if one understands the history of these matters and what settlers regarded here as a destiny manifest for waste land and waste people, these mountains hold an Indian reservation, Fort Belknap, home of the Gros Ventres and Assiniboine Sioux. Once I spent a day there with Mike Fox, a native of the reservation and the tribe's wildlife biologist. It was his job to supervise the herd of bison his people were bringing back to the land. The herd had grown sufficiently to accommodate a few feasts during the year, a practice that had been long abandoned but was being resurrected with these bison. The tribe was coming together again. I'd heard that the tribal courts once sentenced a woman convicted of child neglect to follow the bison and watch the example of cows caring for their calves. Fox took me to the top of a rock outcrop overlooking the herd and showed me petroglyphs, ancient drawings of bison that people of his bloodline had made when they hunted here.

East of Fort Belknap, the mountains peter out onto the Missouri River breaks. The plains here are as harsh and rocky as they come, sagebrush and rattlesnakes, so, naturally enough, this is public land—the Charles M. Russell Wildlife Refuge and adjoining federal holdings managed by the Bureau of Land Management. Like most public lands in the west, it is viciously overgrazed. Some biologists here told me that when their colleagues were commissioned to assess the condition of habitat on a section of this public land, the rancher who leased it (federal land, i.e., your land) chased them off with threats of violence. The rancher's ban stood. Still, the cows leave sections of the land alone, so antelope coexist with their grazing.

I've spent most of my life around our continent's wild ungulates, which in my native Midwest meant white-tailed deer. When I moved to the Rockies, I got to know mule deer and elk. There is a common thread through these species that one quickly senses, despite their differences. The antelope, though, is a different beast altogether, skittish beyond measure and powerfully built. An antelope is to a white-tailed deer what a Porsche is to a Ford Fiesta. This explains something crucial. The other megafauna—the deer, bison, and elk— are interlopers in this land, as are even the bison Mike Fox's ancestors hunted and drew. They came across the land bridge with humans ten thousand years ago to fill the niches left by overkill, the hunting to extinction of all of the big mammals of North America. The antelope is the lone exception. Paul Martin, the paleontologist who first figured out that the overkill had occurred, said antelope survived because they are "gracile," a nice word for what sports fans mean by both "quick" and "fast." Simply, even if those early settlers were bent on hunting antelope to extinction, they would pay hell doing so with stone-tipped spears.

To hunters into our time, they are the most demanding game. One sneaks on hands and knees in this treeless, featureless landscape, hoping to inch within shooting distance of a wary animal, first spotted from two miles away. Even the glint of a forehead above a piece of sagebrush can make them demonstrate that they are indeed the continent's fastest land mammal. The floor of the landscape is constructed mostly of prickly pear cactus and cinder-like rock, so miles on hands and knees seem longer.

At least that's how some of us do it. I'd been told going into my hunting trip that I would see other methods. Modern times have brought those hateful little four- and three-wheeled motorcycles known as ATVs (all-terrain vehicles). "Sportsmen" use these to chase down and kill antelope. Now it is possible to see pickup trucks leave the area after a weekend's hunt with antelope piled up in the back like cordwood. The state is liberal with limits because ranchers hope to keep the number of antelope low. It is axiomatic here that grass is for cows. Everything else is classified as "varmint." This extends downward to prairie dogs. During the off season, these same sportsmen hone their skills by crouching outside a prairie dog town with a high-powered rifle, then vaporizing any dog that pops up. There are even organized contests.

I don't hunt antelope much anymore.

Venery, the art of hunting, may be archaic as a term. As a practice, it is certainly old, but not extinct. When I take to the woods with a rifle each fall, I join a long line of forebears, some of whom I would as soon disavow. The bison hunters we know best from antiquity are not Assiniboine Sioux, but Poles, Germans, and Romans. I say "know best" as a Caucasian. These bison hunters are of my culture, and I feel as if I know them. The written records are good enough to give me some idea of their motivation.

The closest living relative of the American bison is the Polish steppe wisent (pronounce the "w" like a "b" and you have the linguistic connection). They probably both descend from a common ancestor that once wandered Eurasia. The European bison still survives in Bialowieza, a patch of relict forest on the Poland-Lithuania border. In his book *Landscape and Memory*, the historian Simon Schama makes extensive use of the forest as a focal point of European imagination, a stretch of wild that helps define a self-image. Schama's own ancestors are from the region, but forest has served its role through most of European history.

There is, for instance, the case of Hussowski (or by his Latin name, Nicholas Hussovianus), a Polish scholar at Rome in about 1520. Part of a bishop's retinue, he nonetheless traipsed about court in thigh-high boots and a sable coat, a sort of sixteenth-century wildman getup. He was affecting a link to the Sarmatians, half-wild horse-riding hunters of northern Europe, just as a modern writer from Montana living in Manhattan might affect a buckskin jacket and moccasins. He was given to writing long odes to bison; he even dedicated one to Pope Leo X, who was himself a hunter.

The hunting tradition among Europe's elite would not end in the sixteenth century. Notes Schama:

> In Hussowski's prototype of Polish bison lore and in many accounts which followed over the next century, like that of Ritter Sigismund von Herberstein, the Austrian ambassador to Muscovy, the animal was depicted as a miraculous relic of presocial, even prehistoric past—a tribal, arboreal world of hunters and gatherers, at the same time frightening and admirable. The bison became a talisman of survival. For as long as the beast and its succoring forest endured, it was implied, so would the nation's martial vigor. Its very brutishness operated as a test of strength and justice.

Not that Europe itself was a brutish place then or in succeeding years. Even the immediate region around the Polish forest was civilized, in the agricultural sense of the word. The Polish noblemen who hunted the forest were themselves often wheat farmers—or more accurately, lords over wheat farmers—who were exporting their crops to growing urban markets, just as many of the ATV-borne antelope hunters in Montana are off-duty wheat farmers. And the more civilized the place became, the more the rich especially would use the forest to play at being hunter-gatherers. Generations of Polish kings trooped to the forest with full entourages of nobility to hunt bison and elk. Hermann Göring hunted there, and in fact was instrumental in preserving the steppe wisent during World War II. Nikita Khrushchev followed.

Seen in this light, hunting comes to look very much like the rest of the sumptuary practices of European nobility, one more pleasure to which the rich maintain access as the world shrinks to monotony and drudgery for the rest of us. Middle Eastern archaeology contains a hint of this phenomenon. As agriculture took hold in Turkey, there was suddenly a profusion of bronze work—delicate, finely wrought carvings—depicting graceful stags with outsized antlers. One suspects this doesn't signal a resurgence in hunting but a decline.

At bottom, this says not much more than, given the means, people will go hunting. But the same is true of a range of other activities that agriculture has overruled, like eating varied, delicious food. Given the means to be human, we will.

I no longer hunt antelope, but I hunt elk and deer every fall, beginning on the third Sunday in October, when Montana's six-week big game season opens. I am a richly privileged human, in that I get to do so on my own land. It's not much, seventy acres, but it adjoins a rack of similar parcels that themselves back onto public lands extending for miles along mountain ridges. The route to my own private Bialowieza is arduous, the very reason it remains undeveloped. I leave my back door and immediately begin climbing, twenty minutes or so of a hill so steep that sometimes it is easiest to ascend on all fours. All of my neighbors own strips of this land up here, but most have never seen it. It's easy enough to be alone in our overcrowded world if you're willing to walk a bit.

Outsiders think of Montana as a wild, untouched place; it is anything but. It has been logged, plowed, and grazed to the point that it is difficult to find any intact natural plant communities, no matter how it might look to the untrained eye. Here and there, however, up a ridge too steep for cows, sheep, or fences, there are refugia. My land is such a place. At ridgetop it is a long roll of lush native bunchgrasses and ponderosa pine savannah. It lies within sight of a city of a hundred thousand, yet is a pocket of high-quality, low-elevation habitat. Most of this is sheer luck, but some of it is my doing. I've protected this place for a dozen years. I visit it frequently year-round, but always on opening day of big-game season.

On my latest opening day, I wasn't out an hour when I whiffed a shot at an enormous mule deer buck; not jitters so much as a subconscious desire not to let this end so soon. I knew there would be more deer, there always are, and, with great luck, the elk would come.

The appearance of an elk herd here in fall is not rare, but not usual either; they tend to use these grasses when they most need them, in the dead of winter; in fall, they usually frequent higher ground. When it happens, though, it is an overwhelming experience, thirty or so of these animals sweeping over a ridge at a canter, heads high. Which is why I was shocked an hour after missing the deer to hear a bull elk bugle a hundred yards away. I crept and stalked my way toward a clump of trees where this animal was hiding, then crouched down behind a log just in time for the trees to come suddenly alive with a swarm of big, tawny bodies.

No canter, though, they were walking. And suddenly I was surrounded on all sides by cow elk and their calves, some no more than fifty feet away. So I hid, breathing as gently as possible for an hour, watching elk, none of them bulls (i.e., legal game). One cow finally spotted me and announced my presence in what I can only call a bark. The animals stood their ground and watched me as a calf ran from the trees to the barking mother's side. The bull never stepped out of the trees. It grew dark, so I packed up my rifle and headed home. An unsuccessful hunt? Guess again.

So it went, through six weeks of determined hunting. I walked ten pounds off my frame and saw a few more skittish deer, but global warming's balmy autumn kept the animals on their summer grounds and out of my reach.

It would be easy enough to lump my hunting with Göring's and Hussovianus's. I have no sable coat and long boots, but affectation is there nonetheless. When I am sneaking through the woods with a rifle, part of me is still eight years old and wearing the Davy Crockett fake coonskin cap that was standard issue for any fifties kid of my gender. I admit to a certain amount of pathetic romanticist behavior, but at the same time I feel sorry for those who have no romanticism in their lives. Still, at the end of a long season, the freezer was empty and that was real. I like beef but can't bring myself to buy it, knowing what I know. I was staring at a long winter of tofu and beans.

Then, a week or so after the season ended, I heard a rumor of a situation that would allow me to go hunting again, and hunt I did. First, on the Internet. The rumor was that the bison business, being reestablished as a viable substitute for cattle by many Montana ranchers, was having a bad year. Ranchers had overestimated demand and so were overstocked. One could buy a yearling bull for as little as three hundred dollars. The Internet led me to a directory of bison ranchers (www.Hussovianus.com?) and a phone number, and pretty soon I was talking with Paul Daniels at the Heart Bar Heart Ranch, a forty-five-minute drive from my home, up the Blackfoot River. And, no, he hadn't sold any bison yet, but he was thinking about it, and he'd think about selling me one the next day if I showed up. And so I did, on a crisp morning with no snow, and the ground frozen rock solid.

I drove up Montana Highway 200, which weaves through narrow canyons with the Big Blackfoot River. About halfway through the drive, I remembered that the highway followed an old trail used by Lewis and Clark, and by the Salish Indians long before. They called it Cokahlarishkit, which means something like "going for the bison." Every year they used this trail to leave the mountains and hunt bison on the plains. They used ancient rock cairns instead of the Internet to find the way.

A herd of maybe a hundred bison grazed nearby when I drove up to the ranch house. I went in, sat in the glow of the kitchen, talked about the weather for the requisite amount of time, had a cup of coffee. Then Daniels and I climbed into his pickup and drove out among the bison. He pointed one out, and I shot it in the ear. Mrs. Daniels fired up a tractor with a front-end loader, which we used to hoist the animal for dressing it out. A ranch hand joined us and three sets of practiced hands made quick, clean work of gutting the bison. The loader dropped it into my pickup and I drove it home; I used a hand-crank hoist to hang it in a big pine tree in front of my house, where I skinned it. I let the carcass hang there for a day or so, while magpies and Clark's nutcrackers pecked at its edges, scavengers making a living off a bison kill as generations of their ancestors had before; no one needed to tell them dead bison was on hand or how to go about their business.

Then I spent a day butchering the carcass, reducing it to 350 pounds of steaks, roasts, and stew meat, sharpening my knives as I had been taught as a teenager, guiding the clean edge of a tool with enough knowledge to make my own food. Romanticism aside, it's still meat.

NOTES

I am by habit and training a journalist, not a scholar, and so have a journalist's aversion to footnotes. My usual habit is to simply make references clear within the text. Full publication information on those sources can be found in the bibliography. In some cases, that practice was unwieldy, so I have included the notes below.

AROUSAL

13 **the "omnivore's dilemma"** Rozin's idea is discussed in Susan Allport, *The Primal Feast*.

16 **reported by Colin Turnbull** Also discussed in Susan Allport, *The Primal Feast*, as are Lorna Marshall's comments.

19 **Martin argued they were hunted to death** Aside from Martin's own writings, a good summary of the extinctions and the record of human migration to the New World is Brian M. Fagan, *The Great Journey*.

WHY AGRICULTURE?

24 **rice was domesticated** An account of this domestication and the social history it spawned in the New World can be found in Judith A. Carney, *Black Rice* (Cambridge, Mass.: Harvard University Press, 2001).

24 **Israeli scientists . . . found it exceedingly easy** The research is summarized in Bruce Smith, *The Emergence of Agriculture*. The same source is an excellent summary of current research into the roots of domestication in general.

29 **Charles Higham writes** His essay "The Transition to Rice Cultivation in Southeast Asia" appears in T. Douglas Price and Anne Birgitte Gebauer, *Last Hunters, First Farmers*.

31 **Following a specific description in Genesis** Juris Zarins's theory is summarized in Colin Tudge, *Neanderthals, Bandits, and Farmers*.

34 **Cahokia was occupied** Cahokia was a famous city, according to the liter-
ature of early agriculture in the New World. I gathered much detail in a
visit to the place, but a good account of the ancient city can be found in
William H. MacLeish, *The Day Before America*.

36 **They could also grind the grain** T. Molleson, "The Eloquent Bones
of Abu Hureyra." *Scientific American*, vol. 271, no. 2, August 1994,
pp. 70–75.

37 **Mark Cohen ticks off a list** MacLeish quotes him in *The Day Before
America*.

40 **In 2001, Dr. Sarah A. Tishkoff** Reported in Nicholas Wade, "Genetic
Study Dates Malaria to the Advent of Farming," *The New York Times*,
June 22, 2001.

WHY AGRICULTURE SPREAD

48 **"Perhaps the most striking feature"** T. Douglas Price, Anne Birgitte
Gebauer, and Lawrence H. Keeley, "The Spread of Farming into Europe
North of the Alps," in Price and Gebauer, *Last Hunters, First Farmers*.
Much of the detail I use on the spread of the LBK people comes from
their work.

54 **"The introduction of the padded horse collar"** L. T. Evans in his
landmark history, *Feeding the Ten Billion*, a pivotal work and an invaluable
source throughout.

HARD TIMES

68 **The food writer Brian Murton** His essay "Famine" appears in volume II
of *The Cambridge World History of Food*.

70 **a time of "hungry ghosts"** Jasper Becker, *Hungry Ghosts: Mao's Secret
Famine*, provided most of the detail I use from this period.

72 **Amartya Sen, in his book** *Poverty and Famines: An Essay on Entitlement and Deprivations* (Oxford: Clarendon Press, 1981). Quoted in L. T.
Evans, *Feeding the Ten Billion*.

73 **The food historian Sophie Coe** Her book is *America's First Cuisines*.

79 **"bonds of natural affection were loosened"** Larry Zuckerman (*The
Potato*) reports this quote along with the longer one that follows.

81 **As much as spices and slavery** The role of sugar in geopolitics and colonialism is most thoroughly explored in Sidney W. Mintz, *Sweetness and
Power*.

MODERN TIMES

90 **The Wallace family of Iowa** The detail about this remarkable family
comes largely from John C. Culver and John Hyde, *American Dreamer*.

90 **"In 1936 the rains failed again"** Donald N. Duvick, "Responsible Agricultural Technology: Private Industry's Part," *Pro Rege*, June 1990, vol. 26, no. 4, pp. 2–13.

98 **the vast region of the gulf ecologists call the "Dead Zone"** For a thorough study of nitrogen pollution in general, and of the Dead Zone in particular, see the National Research Council's book *Clean Coastal Waters: Understanding and Reducing the Effects of Nutrient Pollution* (Washington, D.C.: National Academy Press, 2000).

A VANGUARD OF FEUDALISM

105 **"Rats and buffaloes in the Punjab"** "Indian Agriculture: Prowling Tiger, Slobbering Dog," *The Economist*, February 17, 2001, p. 46.

TO SEE THE WIZARD

124 **Five companies—Cargill, Incorporated** For statistics on the soybean industry, I rely on John A. Schnittker's paper "The History, Trade and Environmental Consequences of Soybeans in the United States," *World Wildlife Fund*, December 1997. A parallel study on corn by C. Ford Runge and Kimberly Stuart, "The History, Trade and Environmental Consequences of Corn Production in the United States," *World Wildlife Fund*, March 1997, provides data on that commodity.

127 **Today the richest 2 percent** Derives from USDA's *Census of Agriculture*, 1992.

129 **In 1950, farmers grossed forty-one cents** From USDA's *Census of Agriculture*, 1992.

138 **ADM makes most of its corn** The groundbreaking investigation of ADM was James Bovard's report: *Archer Daniels Midland: A Case Study in Corporate Welfare*, Cato Institute, Washington, D.C., September 26, 1995. Much of the detail of the empire, and of the company's political machinations and legal troubles cited in this chapter were first reported there.

WHY WE ARE WHAT WE EAT

151 **Kenneth Albala writes** His essay on the history of foods in southern Europe is on pp. 1203–10 of *The Cambridge World History of Food*, 2000.

156 **Françoise Sabban writes** Her essay on the history of foods in China is on pp. 1165–74 of *The Cambridge World History of Food*.

157 **There may be a rationale behind many taboos** This long list of Hindu food taboos, and many others from other religions, is found in Louis Grivetti's study, "Food Prejudices and Taboos," also in *The Cambridge World History of Food*, pp. 1495–1510.

163 **Robert Pirsig points out that to a hog in a pen** In *Lila: An Inquiry into Morals* (New York: Bantam Books, 1991), Pirsig makes the argument that a hierarchical system has evolved as a sort of species, and that it, not humans, is in control.

166 **Jeffrey Pilcher reports that immigrants** His essay "Food Fads" appears in *The Cambridge World History of Foods*, pp. 1486–94.

166 **"Infused with the spirit"** Betty Fussell, *The Story of Corn* (New York: North Point Press, 1999), p. 265.

170 **"to start chomping their way through the wheat surplus"** Harvey Levenstein, *The Paradox of Plenty*; this fascinating study of America's eating habits provides much of the detail I use here.

179 **"Societies of abundance are tormented"** Fischler, quoted in Levenstein, *The Paradox of Plenty*.

A COUNTERAGRICULTURE

191 **Some of United Fruit's excesses in maintaining that network rivaled those of slavery.** The history of The United Fruit Company in Central America has been written about in great detail. Here are a few sources: Paul Dosal, *Doing Business with the Dictators: A Political History of United Fruit in Guatemala, 1899–1944* (Wilmington, Del.: Scholarly Resources Books, c. 1993); Lester Langley and Thomas Schoonover, *The Banana Men: American Mercenaries and Entrepreneurs in Central America, 1880–1930* (Lexington, Mass.: University of Kentucky Press, 1995); Thomas McCann, *An American Company: The Tragedy of United Fruit* (New York: Crown, 1976); Stephen Schlesinger and Stephen Kinzer, *Bitter Fruit: The Untold Story of the American Coup in Guatemala* (Garden City, New York: Doubleday, 1982).

195 **"It's 6:30 a.m. While most New Yorkers"** *The Christian Science Monitor*, August 29, 2001.

197 **"These internships, which entail working"** Julie Flaherty, "Planting the Seeds for a Resurgence of the Small Farm," *The New York Times*, October 14, 2000.

201 **Fittingly, the Slow Food Movement has a manifesto** The Slow Food Movement's Web site is: http://www.slowfood.com.

I EAT, THEREFORE I KILL

204 **This is Choteau County** Timothy Egan, "Failing Farmers Learn to Profit from Federal Aid," *The New York Times*, December 24, 2000.

BIBLIOGRAPHY

Abram, David. *The Spell of the Sensuous*. New York: Vintage Books, 1997.

Ackerman, Diane. *A Natural History of the Senses*. New York: Vintage Books, 1990.

Allport, Susan. *The Primal Feast: Food, Sex, Foraging, and Love*. New York: Harmony Books, 2000.

Becker, Jasper. *Hungry Ghosts: Mao's Secret Famine*. New York: Henry Holt, 1998.

Bovard, James. *Archer Daniels Midland: A Case Study in Corporate Welfare*. Cato Institute, Washington, D.C., September 26, 1995.

Brody, Hugh. *The Other Side of Eden: Hunters, Farmers, and the Shaping of the World*. New York: North Point Press, 2000.

Cavalli-Sforza, Luigi Luca, and Francesco Cavalli-Sforza. *The Great Human Diasporas: The History of Diversity and Evolution*. Cambridge, Mass.: Perseus Books, 1995.

Coe, Sophie. *America's First Cuisines*. Austin: University of Texas Press, 1994.

Crosby, Alfred. *Ecological Imperialism: The Biological Expansion of Europe, 900–1900*. Cambridge, England: Cambridge University Press, 1986.

Culver, John C., and John Hyde. *American Dreamer: A Life of Henry A. Wallace*. New York: W. W. Norton & Company, 2000.

Diamond, Jared. *The Third Chimpanzee: The Evolution and Future of the Human Animal*. New York: Harper Perennial, 1992.

———. *Guns, Germs, and Steel: The Fates of Human Societies*. New York: W. W. Norton & Company, 1997.

Drache, Hiram M. *Legacy of the Land: Agriculture's Story to the Present*. Danville, Ill.: Interstate Publishers, Inc., 1996.

Evans, L. T. *Feeding the Ten Billion: Plants and Population Growth*. Cambridge, England: Cambridge University Press, 1998.

Fagan, Brian M. *The Great Journey: The Peopling of Ancient America*. London: Thames and Hudson, 1987.

Fernández-Armesto, Felipe. *Civilizations: Culture, Ambition and the Transformation of Nature*. New York: Free Press, 2001.

Fouts, Roger, and Stephen Tukel Mills. *Next of Kin: My Conversations with Chimpanzees*. New York: Avon Books, 1997.

Fussell, Betty. *The Story of Corn: The Myths and History, the Culture and Agriculture, the Art and Science of America's Quintessential Crop*. New York: North Point Press, 1999.

Hanson, Victor Davis. *Fields Without Dreams*. New York: Free Press, 1996.

Heiser, Charles B. *Of Plants and People*. Norman: University of Oklahoma Press, 1985.

Jolly, Alison. *Lucy's Legacy: Sex and Intelligence in Human Evolution*. Cambridge, Mass.: Harvard University Press, 1999.

Kiple, Kenneth F., and Kriemhild Coneé Ornelas, eds. *The Cambridge World History of Food*. Cambridge, England: Cambridge University Press, 2000.

Lasch, Christopher. *The True and Only Heaven: Progress and Its Critics*. New York: W. W. Norton & Company, 1991.

Levenstein, Harvey. *The Paradox of Plenty: A Social History of Eating in Modern America*. New York: Oxford University Press, 1993.

MacLeish, William H. *The Day Before America: Changing the Nature of a Continent*. Boston: Houghton Mifflin Company, 1994.

Matson, Pamela A., Rosamond Naylor, and Ivan Ortiz Monasterio. "Integration of Environmental, Agronomic, and Economic Aspects of Fertilizer Management." *Science*, April 1998, vol. 280, pp. 112–15.

Mintz, Sidney W. *Sweetness and Power: The Place of Sugar in Modern History*. New York: Penguin Books, 1985.

———. *Tasting Food, Tasting Freedom: Excursions into Eating, Culture, and the Past*. Boston: Beacon Press, 1996.

Naylor, Rosamond L., et al. "Nature's Subsidies to Shrimp and Salmon Farming." *Science*, October 1998, vol. 282, pp. 883–84.

———. "Effects of Aquaculture on World Fish Supplies." *Nature*, June 2000, vol. 405, pp. 1017–21.

Nelson, Richard. *Make Prayers to the Raven: A Koyukon View of the Northern Forest*. Chicago: University of Chicago Press, 1983.

———. *The Island Within*. New York: Vintage Books, 1990.

Price, T. Douglas, and Anne Birgitte Gebauer. *Last Hunters, First Farmers*. Santa Fe: School of American Research Press, 1995.

Schama, Simon. *Landscape and Memory*. New York: Vintage Books, 1996.

Schlosser, Eric. *Fast Food Nation: The Dark Side of the All-American Meal*. Boston: Houghton Mifflin Company, 2001.

Smith, Bruce. *The Emergence of Agriculture*. New York: Scientific American Library, 1995.

Tattersall, Ian. *Becoming Human: Evolution and Human Uniqueness*. New York: Harcourt, Brace & Company, 1998.

Tudge, Colin. *Neanderthals, Bandits, and Farmers: How Agriculture Really Began*. New Haven: Yale University Press, 1998.

Wilson, Edward O. *Biophilia*. Cambridge, Mass.: Harvard University Press, 1986.

———. *Consilience: The Unity of Knowledge*. New York: Vintage Books, 1999.

———. *The Diversity of Life*. New York: W. W. Norton & Company, 1992.

World Resources Institute. *Food Consumption and Disruption of the Nitrogen Cycle*. www.wri.org/wri/pdf/critcons.

Worster, Donald. *Rivers of Empire: Water, Aridity, and the Growth of the American West*. New York: Oxford University Press, 1985.

———. *Under Western Skies: Nature and History in the American West*. New York: Oxford University Press, 1992.

———. *The Wealth of Nature*. New York: Oxford University Press, 1993.

Zuckerman, Larry. *The Potato: How the Humble Spud Rescued the Western World*. New York: North Point Press, 1998.

ACKNOWLEDGMENTS

This book is indebted to a lifetime's worth of colleagues and friends who have shared ideas and information and offered support. No one does this sort of thing alone, at least I don't.

Beyond that, though, there are specific contributions that need to be acknowledged here. First, my thinking about agriculture was greatly helped by the John S. Knight fellowship program, which supported me for a year at Stanford University. The reading, thinking, and research I was able to do there in 1994–95 steered me to my conclusion that agriculture is our time's biggest environmental story, and I've been working on it ever since.

Of particular help was a chance encounter there with Rosamond Naylor, now a fellow at Stanford's Center for Environment and Policy. Roz, a valued friend ever since, has been a continuous source of information, challenge, inspiration, and support. In particular, she was the link to the McKnight Foundation, which supported my research and travel in the developing world, work that produced my earlier book *Food's Frontier*. That, in turn, led to a similar project with the Rockefeller Foundation. The support from both foundations was crucial to gathering the detail and rounding out the general direction of these ideas.

The Center for Environment and Policy also sponsored my trip to the Yaqui Valley of Mexico, reported herein. I'm indebted to Roz for that, as well as to Pamela Matson and Walter Falcon for sharing their research and ideas.

My editor, Rebecca Saletan, had an enormous role in shaping this book, especially in helping me rework some wildly disparate ideas into a more coherent flow.

My largest debt, though, on this and all other projects, is owed to Tracy Stone-Manning, my wife. It simply would not be possible without her.

INDEX

aborigines, 14
Abram, David, 6–7, 14
Ackerman, Diane, 150, 155
ADM, *see* Archer Daniels Midland (ADM)
advertising: and processed food, 176–78
A. E. Staley, 147; *see also* Staley, Gene
Africa: Ik people, 16; !Kung people, 17; rice in, 24; and spread of European agriculture, 53–54
Ag Processing, Inc., 124
aggressive exotics, 43
agriculture: arable land limit, 87–89; beginnings, 24–27; catastrophe as precondition, 28–31; commodities as focus, 97–98, 129, 130, 188, 189, 190; and concept of progress, 119, 185–87; and dehumanization, 8, 82, 134; environmental effects of industrialization, 98–101, 108, 109–11; exporting U.S. products, 126–27; and famine, 68–73; vs. farming, 187–88; and floodplains, 28, 29, 31; and government, 72–73, 79; vs. growing food, 187–88, 190; impact of wheat-beef culture, 80–81; impact on health of early settlers, 36–37; and imperialism, 45, 70–71; and irrigation, 101–103, 109, 197; preconditions, 27–31; vs. proto-agriculture, 25, 26–27; reasons for invention,
23–24; relationship to poverty, 33–34, 37, 72, 120–21; role of cultivation, 25; sedentism as precondition, 30, 31; shift from increasing acreage to increasing yield, 86–87, 89–96; spread of domestication, 44–66; tropical vs. temperate, 82; and values, 37–41; and water-quality problems, 99–101, 137–38; and wealth vs. poverty, 33, 120–21; *see also* catastrophic agriculture
Ajimoto, 147
Alaska Highway, 56
Albala, Kenneth, 151, 154
allergies: corn pollen, 142
Allport, Susan, 12, 16, 178–79
alpacas, 61, 77
American Can Company, 177
American Indians, 15, 25, 30, 51, 204, 206
American Maize, 147
American Publishing Co., 146
Andes Mountains: climate for potatoes, 74; Inca civilization, 24; potato as agricultural basis, 76–77; potato domestication, 24, 74
Andre & Cie, 124
Andreas, Dwayne: efficiency joke, 94; interview, 142–46; takes over ADM, 128; *see also* Archer Daniels Midland (ADM)
Andreas, Martin, 142

223

cave art, 9–10, 47, 49
cereals: processed, 165; *see also* grains
CG system, 107
chaos theory, 186
Charles M. Russell Wildlife Refuge, 204–205
chemical fertilizers: consumption growth trend, 109; and crop yield, 91, 93; environmental impact, 98–99, 100; and hybridization, 91; nitrogen in runoff, 99, 137–38; role in industrial agriculture, 128
Cheng Ho, 70
Chez Panisse, 198–200
Chicago Sun-Times, 146
chickpeas, 114
children: and "edible schoolyard" project, 199–200; and obesity, 184; and sugar, 183, 184
chimpanzees: similarities to humans, 11–12; use of food as bonding experience, 15–16
China: early agricultural settlements, 24, 32; early views on food, 155–56; and European trade, 52, 73; famine in, 69, 70–71, 72; as grain producer, 52, 53; Great Leap Forward, 70, 71, 72; Great Wall, 44–45; imperialism in, 70, 71; invention of horse collar, 54; rice domestication, 24; similarity to neo-Europes, 52; Ta-hsi culture, 32
Chiquita bananas, 176
chocolate, 152–53
Christianity: asceticism in, 153, 154–55; body-soul dualism, 153–54; and "daily bread," 159; feasting and fasting, 154; initial role of food, 153
Churchill, Winston, 84
CIMMYT (International Center for the Improvement of Maize and Wheat), 107
Clark, Ralph, 125
Clinton, Bill, 84, 141
Clovis people, 20
coevolution, 38–39, 40–41, 73, 187
Coca-Cola, 183

Coe, Sophie, 73
Cohen, Mark, 37
Colgate & Company, 168
Colorado River, 102, 106
Columbia River, 31, 57, 102
Columbus, Christopher: introduces pigs to New World, 63
Combarelles, *see* cave art
communal feasting, 15–16
commodities: vs. other food, 98, 188, 189, 190
communicable diseases, *see* diseases
companion planting, 192
ConAgra, 147
Conservation Reserve Program, 87
Consultative Group on International Agriculture Research (CG system), 107
Continental Grain Co., 124, 147
Cook, Captain James, 58
cooperatives: farm, 129–30
corn (maize): and ADM, 124, 133–34, 138–40; American corn in Mexico, 133, 161; China as producer, 52; as commodity, 98, 129, 137, 167, 169; as crop in surplus, 97–98, 128, 178; domestication, 24, 45–46; effect of subsidizing, 120–21; environmental impact of production, 100–101; ethanol derivative, 140–42; exporting U.S. product, 126–27; grain-type, 98, 133; hybridization, 89–92, 119; increasing yield, 89–92, 93; as major U.S. crop, 98; in Mayan and Pueblo civilizations, 159; as most important crop, 73; and neo-Europe concept, 52; nixtamalization, 167–68; origin, 24; processing, 167–70; in world's diet, 120–21
corn pollen, 142
Corn Products Company, 168, 169, 177
corn syrup: and ADM, 124, 138–40; early use, 168–69; marketing, 170; vs. sugar, 139; and surplus corn, 98
cornstarch, 168, 169
counterculture, *see* food counterculture